# THE DAY WE WALKED ON THE MOON

## A Photo History of Space Exploration

Book and cover design
by Georgia Morrissey

ISBN  0-590-42760-1
Copyright © 1990 by George Sullivan.
All rights reserved. Published by Scholastic Inc.

12 11 10 9 8 7 6 5 4 3 2 1     0 1 2 3 4 5/9

Printed in the U.S.A.          34

First Scholastic printing, January 1990

Special thanks are due H. Thomas Jaqua and Althea Washington of the National Aeronautics and Space Administration for their interest in this book and their cooperation in providing background information and photographs.

*Edward White, Virgil (Gus) Grissom, and Roger B. Chaffee.*

To Edward White, Virgil (Gus) Grissom, and Roger Chaffee, the three astronauts who died when fire swept their spacecraft during a preliminary check at John F. Kennedy Space Center on January 27, 1967. These three men had been scheduled to fly the first *Apollo* spacecraft.

# THE DAY WE WALKED ON THE MOON

by George Sullivan

A Photo History
of Space Exploration

SCHOLASTIC INC.
New York Toronto London Auckland Sydney

# TABLE OF CONTENTS

Moonwalk.........................................................................8
Dawn of the Space Age ...................................... 12
Pathfinders ............................................................ 14
Manned Space Flights ......................................... 16
Famous Firsts ...................................................... 22
"The *Eagle* Has Landed".................................... 26
Living in Space..................................................... 30
Touchdown, *Columbia* ........................................ 34
Astronaut Training................................................ 38
First U.S. Spacewoman ....................................... 42
Shuttle Missions .................................................. 46
National Tragedy .................................................. 50
Back in Space ...................................................... 54
The Future............................................................ 58
Important Dates in Space .................................... 64
For Further Reading............................................. 66
Index ..................................................................... 68

# Moonwalk

A ghostly figure clad in a bulky white space suit slowly descended the ladder. Having reached the bottom rung, astronaut Neil Armstrong carefully extended his booted left foot, then planted it gently on the fine-grained surface of the moon.

The time was 10:56 P.M. (E.D.T.), July 20, 1969. Then the first human on the moon spoke these words: "That's one small step for man, one giant leap for mankind."

That was it. Human beings had met the moon. A dream of centuries had been realized.

With cautious, shuffling steps, Armstrong began moving about on the lunar soil. "The surface is fine and powdery," he said. "It adheres in fine layers, like powdered charcoal, to the soles and sides of my foot. I can see the footprints of my boots and the treads in the fine, sandy particles."

Then Armstrong walked carefully across the surface. He found he could move easily despite his heavy space suit and backpack because the gravity of the moon is only one sixth as strong as that of the earth.

After nineteen minutes, Armstrong was joined outside the landing craft by Edwin "Buzz" Aldrin, Jr. Then, gaining confidence with every step, the two began bounding across the barren landscape, at times even hopping like kangaroos. TV cameras they had set up enabled hundreds of millions of people on Earth to watch.

Meanwhile, the third member of the crew, Michael Collins, piloted the mission command ship in lunar orbit about 70 miles above the surface. When the two explorers had completed their work, they would rejoin Collins for the trip back to Earth.

The astronauts took photos of the

*An astronaut boot-print on the lunar soil.*

# The New York Times

LATE CITY EDITION

Weather: Rain, warm today; clear
tonight. Sunny, pleasant tomorrow.
Temp. range: today 80-66; Sunday
71-64. Temp.-Hum. Index yesterday
69. Complete U.S. report on P. 30.

VOL.CXVIII..No.40,721    © 1969 The New York Times Company    NEW YORK, MONDAY, JULY 21, 1969    11 CENTS

# MEN WALK ON MOON

## ASTRONAUTS LAND ON PLAIN; COLLECT ROCKS, PLANT FLAG

### Voice From Moon: 'Eagle Has Landed'

EAGLE (the lunar module): Houston, Tranquility
Base here. The Eagle has landed.

HOUSTON: Roger, Tranquility, we copy you on the
ground. You've got a bunch of guys about to turn blue.
We're breathing again. Thanks a lot.

TRANQUILITY BASE: Thank you.

HOUSTON: You're looking good here.

TRANQUILITY BASE: A very smooth touchdown.

HOUSTON: Eagle, you are stay for T1. [The first
step in the lunar operation.] Over.

TRANQUILITY BASE: Roger. Stay for T1.

HOUSTON: Roger and we see you venting the ox.

TRANQUILITY BASE: Roger.

COLUMBIA (the command and service module):
How do you read me?

HOUSTON: Columbia, he has landed Tranquility
Base. Eagle is at Tranquility. I read you five by.
Over.

COLUMBIA: Yes, I heard the whole thing.

HOUSTON: Well, it's a good show.

COLUMBIA: Fantastic.

TRANQUILITY BASE: I'll second that.

APOLLO CONTROL: The next major stay-no stay
will be for the T2 event. That is at 21 minutes 26 seconds after initiation of power descent.

COLUMBIA: Up telemetry command reset to re-
acquire on high gain.

HOUSTON: Copy. Out.

APOLLO CONTROL: We have an unofficial time for
that touchdown of 102 hours, 45 minutes, 42 seconds
and we will update that.

HOUSTON: Eagle, you loaded R2 wrong. We
want 10254.

TRANQUILITY BASE: Roger. Do you want the hori-
zontal 55 15.2?

HOUSTON: That's affirmative.

APOLLO CONTROL: We're now less than four minutes from our next stay-no stay, it will be for one complete ... ... ... of the command module.

One of the first things that Armstrong and Aldrin
will do after getting their next stay-no stay will be to
remove their helmets and gloves.

HOUSTON: Eagle, you are stay for T2. Over.

Continued on Page 4, Col. 1

### VOYAGE TO THE MOON

By ARCHIBALD MacLEISH

Presence among us,

            wanderer in our skies,

dazzle of silver in our leaves and on our
waters silver,

            O

silver evasion in our farthest thought—
"the visiting moon" . . . "the glimpses of the moon" . . .

and we have touched you!

            From the first of time,
before the first of time, before the
first men tasted time, we thought of you.
You were a wonder to us, unattainable,
a longing past the reach of longing,
a light beyond our light, our lives—perhaps
a meaning to us . . .

            Now

our hands have touched you in your depth of night.

Three days and three nights we journeyed,
steered by farthest stars, climbed outward,
crossed the invisible tide-rip where the floating dust
falls one way or the other in the void between,
followed that other down, encountered
cold, faced death—unfathomable emptiness . . .

Then, the fourth day evening, we descended,
made fast, set foot at dawn upon your beaches,
sifted between our fingers your cold sand.

We stand here in the dusk, the cold, the silence . . .

and here, as at the first of time, we lift our heads.
Over us, more beautiful than the moon, a
moon, a wonder to us, unattainable,
a longing past the reach of longing,
a light beyond our light, our lives—perhaps
a meaning to us . . .

            O, a meaning!

over us on these silent beaches the bright
earth,

            presence among us

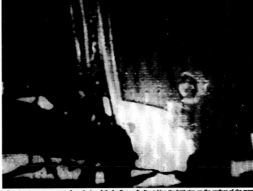

Neil A. Armstrong moves away from the leg of the landing craft after taking the first step on the surface of the moon.

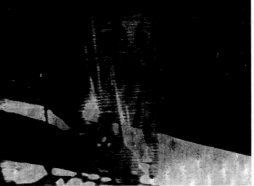

The New York Times from C.B.S. News

Col. Edwin E. Aldrin Jr. climbing down the ladder. The television camera was attached to a side of the lunar module.

Associated Press

Mr. Armstrong, right, and Colonel Aldrin raise the U.S. flag. A metal rod at right angles to the mast keeps flag unfurled.

### A Powdery Surface Is Closely Explored

By JOHN NOBLE WILFORD
Special to The New York Times

HOUSTON, Monday, July 21—Men have landed and
walked on the moon.

Two Americans, astronauts of Apollo 11, steered their
fragile four-legged lunar module safely and smoothly to
the historic landing yesterday at 4:17:40 P.M., Eastern day-
light time.

Neil A. Armstrong, the 38-year-old civilian commander,
radioed to earth and the mission control room here:

"Houston, Tranquility Base here. The Eagle has landed."

The first men to reach the moon—Mr. Armstrong and
his co-pilot, Col. Edwin E. Aldrin Jr. of the Air Force—
brought their ship to rest on a level, rock-strewn plain near
the southwestern shore of the arid Sea of Tranquility.

About six and a half hours later, Mr. Armstrong opened
the landing craft's hatch, stepped slowly down the ladder
and declared as he planted the first human footprint on
the lunar crust:

"That's one small step for man, one giant leap for
mankind."

His first step on the moon came at 10:56:20 P.M., as
a television camera outside the craft transmitted his every
move to an awed and excited audience of hundreds of
millions of people on earth.

#### Tentative Steps Test Soil

Mr. Armstrong's initial steps were tentative tests of
the lunar soil's firmness and of his ability to move about
easily in his bulky white spacesuit and backpacks and under
the influence of lunar gravity, which is one-sixth that of the
earth.

"The surface is fine and powdery," the astronaut reported. "I can pick it up loosely with my toe. It does adhere
in fine layers like powdered charcoal to the sole and sides
of my boots. I only go in a small fraction of an inch, maybe
an eighth of an inch. But I can see the footprints of my
boots in the treads in the fine sandy particles."

After 19 minutes of Mr. Armstrong's testing, Colonel
Aldrin joined him outside the craft.

The two men got busy setting up another television
camera out from the lunar module, planting an American
flag into the ground, scooping up soil and rock samples,
deploying scientific experiments and hopping and loping
about in a demonstration of their lunar agility.

They found walking and working on the moon less
taxing than had been forecast. Mr. Armstrong once reported he was "very comfortable."

And people back on earth found the black-and-white television pictures of the bug-shaped lunar module and the men
tramping about it so sharp and clear as to seem unreal, more
like a toy and toy-like figures than human beings on the
most daring and far-reaching expedition thus far undertaken.

#### Nixon Telephones Congratulations

During one break in the astronauts' work, President Nixon
congratulated them from the White House in what, he said,
"certainly has to be the most historic telephone call ever
made."

"Because of what you have done," the President told the
astronauts, "the heavens have become a part of man's world.
And as you talk to us from the Sea of Tranquility it requires
us to redouble our efforts to bring peace and tranquility to
earth.

"For one priceless moment in the whole history of man all
the people on this earth are truly one—one in their pride in
what you have done and one in our prayers that you will
return safely to earth."

Mr. Armstrong replied:

"Thank you Mr. President. It's a great honor and
privilege for us to be here representing not only the United
States but men of peace of all nations, men with interests
and a curiosity and men with a vision for the future."

Mr. Armstrong and Colonel Aldrin returned to their
landing craft and closed the hatch at 1:12 A.M., 2 hours 21
minutes after opening the hatch on the moon. While the
third member of the crew, Lieut. Col. Michael Collins of the
Air Force, kept his orbital vigil overhead in the command
ship, the two moon explorers settled down to sleep.

Outside their vehicle the astronauts had found a bleak

Continued on Page 2, Col. 1

### Today's 4-Part Issue of The Times

This morning's issue of The
New York Times is divided into
four parts. The first part is de-
voted to news of Apollo 11
and includes Editorials and letters to the Editor (Page 16).
Poems on the landing on the
moon appear on Page 17.

General news begins on the
first page of the second part.
The News Summary and Index
is on the first page of the third
part, which includes sports
news, obituaries (Page 51) and
transportation news and weather reports (Pages 50 and 32).

Financial and business news
begins on the first page of the
fourth part.

Following is the News Index
for today's issue:

|                | Page |            | Page |
|----------------|------|------------|------|
| Bills in Washington | 16 | Food | 42 |
| Books | 33 | Movies | 38-41 |
| Bridge | 37 | Music | 38-41 |
| Business | 53-56 | Obituaries | 51 |
| Buyer | 49 | Society | 37 |
| Chess | 37 | Sports | 43-47 |
| Crossword | 33 | Theaters | 38-41 |
| Editorials | 16 | Transportation | 30, 32 |
| Fashions | 42 | TV and Radio | 55 |
| Financial | 53-56 | Weather | 30 |

News Summary and Index, Page 35

*Astronaut Edwin Aldrin walks on the surface of the moon near the leg of the landing craft. Note footprints in foreground.*

lunar landscape (some of which are reproduced here) and had several chores to perform. Gathering samples of rocks and soil was the most important. They scooped up almost 50 pounds of surface material, putting it in sealed containers for the return voyage.

In all, Armstrong and Aldrin spent two hours and thirty-one minutes on the moon.

Before they climbed back into their landing craft, they planted an American flag in the moon soil. But Armstrong, Aldrin, and Collins were not merely heroes of the nation. They captured the imagination of the entire world. Facing incredible risks, they had helped to launch humanity's greatest adventure — unlocking the mysteries of outer space.

Apollo 11 *crew members (left to right) Michael Collins, Neil Armstrong, and Edwin Aldrin.*

# Dawn of the Space Age

The space age began about twelve years before the first human walked on the moon. On October 4, 1957, the Soviet Union launched the first artificial satellite to circle the earth. It was named *Sputnik*, the Russian word for traveler.

Not quite as big as a basketball, *Sputnik* had a radio transmitter inside that sent out a steady *beep . . . beep . . . beep.*

*Sputnik* was made of aluminum alloys and launched by a military rocket. Once in orbit, it circled the earth every ninety-five minutes at a speed of 18,000 miles per hour.

After making 326 revolutions of the earth, *Sputnik*'s batteries died and it stopped beeping. The satellite stayed in orbit until January 4, 1958, when it fell to earth like a flaming meteorite.

*Sputnik* surprised the world. People were startled by the idea of space travel. It was something out of science fiction. Words such as "astronaut," "orbit," and "satellite" had hardly even been heard.

The Soviet Union launched nine more *Sputnik*s. All were designed to gather information for manned space flight. The last *Sputnik* sent into space was in 1961.

Space travel is common nowadays. Scarcely a month passes without the

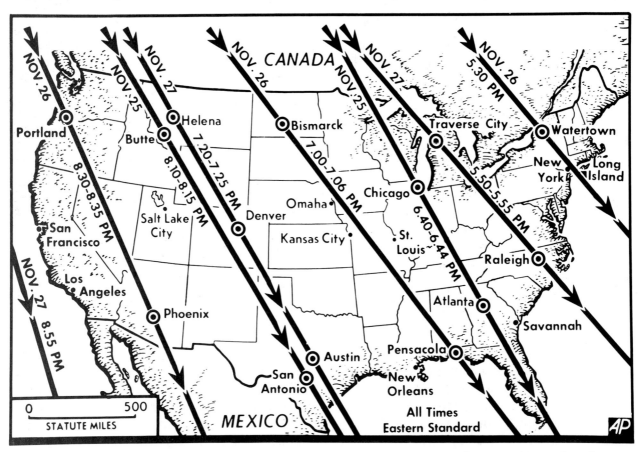

*Black lines indicate the path of* Sputnik *as it passed over the United States on November 25, 26, and 27, 1957.*

*Soviet citizens view a model of* Sputnik *in the Moscow Exhibition Hall.*

launching of a space shuttle, a reusable space vehicle. Space station laboratories, platforms that are used to study the stars, the sun, and the earth's upper atmosphere, have become a reality.

As for satellites, they circle the earth in growing numbers. They bring us television programs from the other side of the world. They enable us to talk by telephone to distant places. They help scientists to map the earth and they gather information about the weather.

There are more than 3,000 satellites spinning around the earth today. *Sputnik* was the first.

# Pathfinders

Space, which is said to begin about 100 miles above the earth, is a hostile environment. Space is silent and airless. The sky is always black and the stars always shine.

Space is subject to extreme heat and extreme cold, and is crisscrossed by belts of dangerous radiation.

Before humans could venture into orbital flight, space suits had to be developed that would keep temperature and air pressure at proper levels.

Spacecraft had to be built that would protect astronauts from high levels of radiation and from collisions with meteors. The spacecraft also had to provide for basic human needs — breathing, eating, drinking, and sleeping.

Overcoming the pull of gravity was another big problem that scientists faced in seeking to climb into space. To lift a spacecraft into the air, rockets and powerful rocket engines had to be developed.

When scientists and engineers finally completed work on the first space vehicles, they tested them, not with humans, but with small animals. On November 2, 1957, the Soviet Union launched a dog named Laika into orbit. Laika was wired to send back data on conditions in space.

The first American space traveler, sent into orbit late in 1959 as part of the Project Mercury program, was a 37-pound, four-year-old chimpanzee named Ham.

There is no sensation of gravity in space. Ham floated in a weightless condition inside the capsule for seven minutes — and survived in good condition.

Not only that, Ham carried out various tasks he had been assigned. For example, Ham was trained to respond to light signals by pulling a lever. When Ham responded correctly, he was rewarded with banana pellets.

*Laika, the Soviet Union's space dog, posed for this photograph before she was sent into the upper atmosphere in November 1957.*

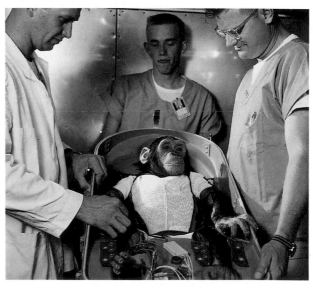

*Ham is prepared for his flight to test the* Mercury *spacecraft.*

At the same time these experiments were being conducted, both the United States and the Soviet Union were sending unmanned spacecraft to the moon, the planets, and around the sun. Little by little, the way was being prepared for manned space travel.

# The New York Times.

LATE CITY EDITION
U.S. Weather Bureau Report (Page 95) forecast:
Rain early today; cloudy later;
rain late tonight and tomorrow.
Temp. range: 65—55. Yesterday: 62.0—55.2.

NEWS SUMMARY AND INDEX, PAGE 95

VOL. CVII—No. 36,443. © 1957, by The New York Times Company Times Square, New York 36, N. Y.

NEW YORK, SUNDAY, NOVEMBER 3, 1957.

25c beyond 100-mile zone from New York City

SECTION ONE

TWENTY-FIVE CENTS

# SOVIET FIRES NEW SATELLITE, CARRYING DOG; HALF-TON SPHERE IS REPORTED 900 MILES UP

## Zhukov Ousted From Party Jobs; Konev Condemns Him

---

### MEYNER'S VICTORY IS SEEN IN SURVEY OF JERSEY VOTERS

**Democratic Governor Likely to Win Re-election Over Senator Forbes Tuesday**

#### A Times Team Report

*A team of New York Times reporters has just completed a survey of political trends and issues in New Jersey. Reports on the election campaign there come from George Cable Wright, Milton Honig, Alfred E. Clark, Leonard Buder, John W. Slocum and Laymond Robinson.*

**By GEORGE CABLE WRIGHT**

The curtain will descend tomorrow night on the New Jersey gubernatorial campaign. The contest—on the surface, at least—appears to have been enacted before a relatively bored audience.

Neither Gov. Robert B. Meyner, the Democratic incumbent, nor State Senator Malcolm S. Forbes, the Republican candidate, has exhibited the ability to rouse the voting public markedly from its apparent apathy.

Beyond the Hudson and the Delaware, however, far greater interest is being manifested in the contest.

The Eisenhower Administration has staked its prestige on the results of the balloting as never before in a state-wide race. Republicans and Democrats alike at the national level are eagerly awaiting the vote tally. Each party hopes to gain from it a trend in its favor.

**Surprise Possible**

The apparent lack of interest locally may well be misleading. It is not an uncommon trait of the state's electorate, as witness 1953, 1954 and 1956. In those years the pre-election tempo turned out to be a "sleeper." The voters, from Cape May to High Point, set their alarms for election morn and flocked to the polls.

As the present campaign progressed, it became increasingly evident that, in all probability, it would be decided on the basis of personality rather than on issues. This was verified by a team of New York Times reporters in the field.

On the basis of findings of The Times' survey team, victory for Mr. Meyner is definitely in-

Continued on Page 60, Column 6

#### DEMOCRATS COUNT ON PARTY VICTORY

**Believe Wagner Can Win Without Liberal Votes in Mayoral Race**

**By LEO EGAN**

Democratic leaders were counting confidently yesterday on obtaining enough votes on their party's line alone to insure the re-election of Mayor Wagner and his running mates next Tuesday. If they can do so it will be the first time since 1933 that a Democrat has received a majority of all the votes in a New York City Mayoral election.

Four years ago Mr. Wagner won by virtue of a split in the opposition between Harold Riegelman, Republican, and Rudolph Halley, Liberal and Independent. Mr. Wagner received just over 45 per cent of the total vote cast.

This year the Liberal party is backing Mayor Wagner and his two city-wide running mates, Controller Lawrence E. Gerosa and City Council President Abe Stark. But Democratic leaders would like to be able to say they could have won without the Liberal endorsement.

Alex Rose, state vice chairman of the Liberal party, referred to this Democratic attitude yesterday in appealing for a large Wagner-Gerosa-Stark vote on the Liberal party line.

"A large vote on the Liberal line is a vote with a special message to the city administration to be independent and to be best guarantee for a clean and effective administration on all levels of city government," he said.

"A large vote on the Liberal line will continue the Liberal party as the political conscience

Continued on Page 53, Column 3

#### Major Sports News

**FOOTBALL**

Navy beat Notre Dame in the nation's top college contest yesterday. Scores of leading games:

| | | | |
|---|---|---|---|
| Alabama | 14 | Georgia | 13 |
| Amherst | 19 | Tufts | 0 |
| Army | 53 | Colgate | 7 |
| Auburn | 13 | Florida | 0 |
| Cornell | 8 | Columbia | 0 |
| Dartmouth | 14 | Yale | 14 |
| Delaware | 23 | Rutgers | 19 |
| Georgia | 13 | Duke | 0 |
| Harvard | 13 | Penn | 6 |
| Iowa | 21 | Michigan | 21 |
| Michigan St. | 21 | Wisconsin | 7 |
| Minnesota | 34 | Indiana | 0 |
| Missouri | 9 | Colorado | 6 |
| Navy | 20 | Notre Dame | 6 |
| N. C. State | 19 | Wake Forest | 0 |
| Ohio State | 47 | Northwestern | 6 |
| Oklahoma | 13 | Kansas St. | 0 |
| Oregon | 37 | Stanford | 26 |
| Oregon St. | 39 | Wash. St. | 25 |
| Penn St. | 27 | W. Virginia | 6 |
| Princeton | 7 | Brown | 0 |
| Purdue | 21 | Illinois | 6 |
| Syracuse | 24 | Pittsburgh | 21 |
| T. C. U. | 19 | Baylor | 0 |
| Tennessee | 35 | N. Carolina | 6 |
| Texas A.&M. | 7 | Arkansas | 6 |
| Vanderbilt | 7 | L. S. U. | 0 |

**HORSE RACING**

Eddie Schmidt won the $86,900 Gallant Fox Handicap at Jamaica by half a length. Bold Ruler was first in the Benjamin Franklin Handicap.

**HOCKEY**

The Rangers routed the Boston Bruins, 5—0.

*Details in Section 5.*

---

### President and Class Honor Academy

The President drinks from fountain he and other members of 1915 class gave to academy. Mrs. Eisenhower watches.

**By W. H. LAWRENCE**
Special to The New York Times.

WEST POINT, N. Y., Nov. 2 — President Eisenhower watched Army defeat Colgate today as the climax to a nostalgic reunion with his 1915 Military Academy classmates. Like any other old grad, the President leaped to his feet and cheered whenever Army threatened or scored—and he had many opportunities this afternoon as the West Point

Continued on Page 45, Column 1

#### Voters Will Settle 7 State Questions; Issues Are Listed

ALBANY, Nov. 2—Voters who go to the polls on Tuesday

the proposed amendments to the State Constitution and whether a constitutional convention should be held.

If performance runs true, only about half those voting will bother to answer the seven questions across the top of every ballot.

The type on the ballot is small and the questions do not always express in the limited space the impact of the proposition.

Following is a description of each proposal, what it would do and the arguments for and against it:

The ballot asks:
"Shall there be a convention to revise the Constitution and amend the same?"

Approval would mean the voters would elect delegates on a party basis in 1958 and those elected would hold a convention the following year, probably in the summer. The convention

Continued on Page 62, Column 4

#### British and French, a Year After, Say Suez Invasion Was Justified

**London Reconciled**

**By DREW MIDDLETON**

LONDON, Nov. 2—In the view of some of those who planned the British-French invasion of Egypt, the situation obtaining in the Middle East a year later justifies that attempt to halt the march of Arab nationalism and its ally, Soviet communism, in the area.

A year ago the Soviet Union had one client and ally in the Middle East, Egypt. Today it has two, Egypt and Syria. The withdrawal of the British and French forces from Suez at the demand of the United Nations has been interpreted by Arab nationalism as a victory and has created a power vacuum into which Soviet imperialism has moved, it is noted.

The view that the invasion

Continued on Page 33, Column 3

**Paris Still Bitter**

**By ROBERT C. DOTY**

PARIS, Nov. 2—The weekend marks the first anniversary of the British-French invasion of Egypt finds most Frenchmen, including those who planned the action, convinced that it was a good idea.

There is no tendency here to push such an idea aggressively. On the contrary, French high officials seem to liquidate as speedily and unobtrusively as possible the remaining economic, political and diplomatic consequences of last fall's events. This is regarded as the logical prerequisite to a restoration of complete interallied confidence and effective action to repair the Western position in th Middle East.

The view here that the invasion Furthermore, the French

Continued on Page 33, Column 2

*This section consists of 136 pages divided into three parts. The news summary and the index will be found on Page 95. Society news begins on Page 90 and obituary articles will be found on Pages 88 and 89.*

---

### A.F.L.-C.I.O. TARGET RESIGNS AS CHIEF OF TEXTILE UNION

**Valente Voices Hope Group Will Stay in Federation —2 More Actions Taken**

Special to The New York Times.

WASHINGTON, Nov. 2—The president of the United Textile Workers, Anthony Valente, resigned today. He said he was acting to help his union retain its membership in the American Federation of Labor and Congress of Industrial Organizations.

The 44,000-member union was one of three cited for corruption last month by the executive council of the parent labor organization. Mr. Valente was declared ineligible to hold office.

Tonight, the union accepted the resignation of Mr. Valente and announced other steps to conform with demands by the A. F. L.-C. I. O. to "clean up" the textile workers operations.

Francis Schaufenbil, secretary-treasurer of the union, told reporters at the end of an all-day meeting of the organization's executive board that "we certainly hope" the actions taken would keep the textile workers in the parent body.

**2 Other Measures**

The board meeting was called today to answer charges brought by the A. F. L.-C. I. O. council.

In addition to accepting Mr. Valente's resignation, the board took these two actions to comply with the council's demands:

1. It agreed to call a special convention "as soon as possible" to elect new officers. The session will be in Washington, New York or Philadelphia.

2. The board "rescinded" a $104,000 severance pay deal for Lloyd Klenert, resigned

Continued on Page 44, Column 3

---

### ZHUKOV HUMBLED

**He Admits 'Mistakes' —Accused of 'Cult' in Armed Forces**

*Text of Soviet communiqué is printed on Page 4.*

**By WILLIAM J. JORDEN**

MOSCOW, Sunday, Nov. 3—Marshal Georgi K. Zhukov, dismissed a week ago as Defense Minister of the Soviet Union, has been removed from all his top posts in the Soviet Communist party.

The party's Central Committee announced last night that Marshal Zhukov had lost his place on the party's central policy-making group, the Presidium, as well as on the Central Committee itself. The principal charge against the hero of World War II was that he had tried to eliminate the Communist party's direction and control of the Soviet armed forces.

The Communist party newspaper Pravda reported this morning that Marshal Zhukov had admitted his "mistakes" during the Central Committee meeting at which he was expelled from the party leadership.

**Anti-Stalin Phrase Used**

He tempered that acceptance somewhat by telling his party comrades that he accepted their criticism of him as being "in the main correct." He also was said to have accepted the admission on his leadership of the armed forces as being of "comradely party assistance to me personally and to other military workers."

The barrel-chested, square-jawed soldier was charged with promoting his own "cult of personality" in the army. This is the phrase used here in reference to Stalin's one-man rule, which was vigorously condemned by the Twentieth Congress of the Communist party last year.

With the help of sycophants and flatterers, the Central Committee said, he was praised to the sky in lectures and reports, in articles, films and pamphlets, and his person and role in the Great Patriotic War [World War II] were overglorified.

The result, the Central Committee charged, was that the whole history of the war had been "distorted." It said that by building himself up Marshal Zhukov had belittled the efforts of the Soviet people, of the

Continued on Page 3, Column 1

---

### Marshal Is Linked to Stalin In Blame for '41 Reverses

**Konev Charges Ex-Chief Distorted History to Create Hero's Role**

Special to The New York Times.

MOSCOW, Sunday, Nov. 3—Marshal Ivan S. Konev, long companion and subordinate of Marshal Georgi K. Zhukov, condemned the former Defense Minister today for "errors in military science."

Marshal Konev's attack was the first derogatory statement leveled against Marshal Zhukov on military grounds.

Soviet commander of the Warsaw Pact forces, Marshal Konev issued his condemnation in an article in today's Pravda, the Communist party organ.

The Konev article said that Marshal Zhukov was responsible along with Stalin for lack of preparedness in the Soviet Union to meet the imminent German attack in June, 1941. It belittled Marshal Zhukov's role in the victories at Stalingrad and Berlin and accused Marshal Zhukov of undue pride and of twisting historical fact.

Continued on Page 3, Column 4

Marshal Ivan S. Konev

#### SOVIET 'STRESSES' SEEN BY THE U. S.

**Washington Expects Strain Behind the Iron Curtain From Zhukov Disgrace**

*State Department statement will be found on Page 6.*

**By RUSSELL BAKER**

WASHINGTON, Nov. 2—The State Department said tonight that the downgrading of Marshal Zhukov's "disgrace" followed only by a short time the expressed desire of Nikita S. Khrushchev, first Secretary of the Soviet Communist party, to send the military leader on a special mission to the United States.

The department said this following so closely "similar action against" other one-time Soviet leaders, demonstrated the polit-

Continued on Page 5, Column 7

#### SATELLITE SIGNAL RECEIVED AT M.I.T.

**Scientists Believe That Orbit Repeats First Sphere— Trackers Are Alerted**

**By The United Press**

CAMBRIDGE, Mass., Sunday, Nov. 3—The first American pick-up of the new Soviet satellite's radio signal was reported early today by the Smithsonian Astrophysical Observatory.

Leon Campbell at the observatory staff said that the satellite apparently was in roughly the same path of 60 degrees as the first Soviet satellite.

Dr. J. Allen Hynek at the observatory, said that 140 miles, the same as the first satellite. Dr. Hynek said. "It seems like a repetition of the orbit of the first satellite," he added.

The Soviet launching of the

Continued on Page 26, Column 4

---

### ORBIT COMPLETED

**Animal Still Is Alive, Sealed in Satellite, Moscow Thinks**

By The Associated Press

LONDON, Sunday, Nov. 3—The Soviet Union announced today it had launched a second space satellite—this one carrying a dog. Radio signals indicated that the animal was living, the Russians said.

A satellite six times as heavy as the one sent up Oct. 4 now is circling the earth every hour and forty-two minutes at a height of 937 miles, Moscow said. This means that the speed is nearly 18,000 miles an hour for the 1,110-pound satellite.

The dog was reported hermetically sealed in a container equipped with an air-conditioning system.

Moscow Radio said data received from the satellite indicated the "functioning of scientific instruments and control of the living activities of the animal are taking place normally."

**First Trip Reported**

The new satellite carries transmitting equipment and apparatus for measuring cosmic rays, temperature and pressure. It also carries equipment for reporting the condition of the dog.

It first passed over the Soviet capital at 11:20 P. M. Eastern Standard Time last night and then completed its first trip around the earth over Moscow at 1:05 A. M. today, the Soviet Union reported.

The announcement said the second satellite was "dedicated to the fortieth anniversary of the great October revolution," which the Communist world will celebrate in Moscow beginning next Thursday.

The new earth satellite is completing its orbit in about seven minutes more than the original Sputnik, still circling the earth.

**Japan Receives Signals**

Moscow said the second satellite was sending out two radio signals.

One, like the "beep" transmitted by the first satellite, is on a frequency of 20,005 megacycles. The other signal, at 40.002 megacycles, is a continuous one.

In Tokyo the Japan Broadcasting Corporation announced that radio signals from the second satellite were being heard.

The corporation picked up the signals twenty-three minutes after Moscow's announcement. The "beep" was at intervals of three-tenths of a second.

A three-stage rocket shoved the original satellite into its orbit. The first Moscow announcement of the second sphere did not explain how it had been sent up.

Although the announcement of the satellite's passing over Moscow indicated an interval of one hour and forty-five minutes

Continued on Page 26, Column 2

#### Mao Is in Moscow; He Hails Soviet Tie

**By MAX FRANKEL**
Special to The New York Times.

MOSCOW, Nov. 2—Mao Tse-tung, leader of Communist China, arrived in Moscow today. He is probably the most important of the gathering here to show the unity and might of international communism.

Virtually all the reigning heads of Communist nations and parties, with the notable exception of President Tito of Yugoslavia, will make the pilgrimage here to join in next week's celebrations of the fortieth anniversary of the Bolshevik Revolution.

Expected in addition to Mr. Mao, who is the Chinese Communist chief of state and party chairman, are Poland's party leader, Wladyslaw Gomulka, and Premier Josef Cyrankiewicz; Premier Janos Kadar of Hun-

Continued on Page 8, Column 3

---

CHINESE COMMUNIST LEADER GREETED IN MOSCOW: Mao Tse-tung, left, the chief of state and Communist party chief, as he arrived yesterday at the capital airport. With him were Nikita S. Khrushchev, center, Soviet Communist chief, and Premier Nikolai A. Bulganin.

# Manned Space Flights

Like the first satellite, the first human to travel in space was a Russian. His name was Yuri Gagarin. An Army major, the 27-year-old Gagarin blasted off from the Soviet Union in his *Vostok 1* spacecraft on April 12, 1961.

Before reentering the atmosphere, *Vostok 1* made one full orbit of the earth, while traveling at 18,000 miles an hour, about 200 miles high. Gagarin was ejected from the spacecraft after reentry, and he parachuted safely to earth. Bigger

Vostok 1, *the world's first manned spaceship, was put on exhibition in Moscow in 1965.*

*Major Yuri Gagarin, the first human to travel in space, waves to the crowd during a visit to England in 1961.*

chutes opened on *Vostok 1*, carrying the spacecraft to a gentle landing.

"I could not see as well as from an airplane, but I could see very well," Gagarin told reporters after the flight. "I saw the spherical shape of the earth. I must say the view of the horizon is unusual and very beautiful. I could see the unusual transition from the light surface of the earth to the blackness of the sky."

Soviet leader Nikita Khrushchev had high praise for Gagarin. "You have made yourself immortal because you are the first man in space," said Khrushchev. Then he added, "Now let other countries try to catch us."

From blast-off to landing, Gagarin's mission lasted 108 minutes. The famed Soviet astronaut died in an airplane crash in 1968.

The United States sought to get back into the race with Project Mercury, which was to put the first American into space. The astronaut chosen for the mission was Alan B. Shepard, a 37-year-old Navy commander.

Twenty-three days after Yuri Gagarin's historic flight, Shepard's *Freedom 7* capsule was launched from Florida's Cape Canaveral. It was not intended to be an orbital flight, as Yuri Gagarin's had been, but merely a short lob.

*Nine feet high and 72 inches wide at its base,* Mercury *spacecraft carried the first Americans into space.*

Shepard stayed in space for only fifteen minutes, reaching a height of 115 miles. *Freedom 7* traveled 302 miles downrange from the Cape. Recovery operations were perfect. "The only complaint I have," Shepard said afterward, "was the flight wasn't long enough."

At Cape Canaveral on February 20, 1962, at two hours and twenty minutes after midnight, Marine Corps Major John Glenn was awakened. He showered, dressed, and sat down to a breakfast of scrambled eggs, steak, toast, orange juice, and coffee.

Then the man who was chosen to be the first American to orbit the earth got into his pressurized space suit and walked out to the launching pad to enter his *Friendship 7* spacecraft and begin the long countdown leading to lift-off.

A television audience of 100 million heard the voice of Mercury control, John "Shorty" Powers, declare, "Glenn reports all spacecraft systems go! Mercury control is go!" The huge Atlas rocket carrying *Friendship 7* gave off a tremendous roar and slowly lifted into the air. "A beautiful sight, looking eastward across the Atlantic," Glenn said.

*Alan Shepard, the first American in space, is successfully recovered after ocean splashdown.*

"All the News That's Fit to Print"

# The New York Times.

LATE CITY EDITION
U. S. Weather Bureau Report (Page 95) forecasts Increasing cloudiness today. Snow, rain tonight. Rain tomorrow. Temp. range: 38—26; yesterday: 37—33.

VOL. CXI . No. 38,014. | NEW YORK, WEDNESDAY, FEBRUARY 21, 1962. | 10 cents beyond 50-mile zone from New York City except on Long Island. Higher in air delivery cities. | FIVE CENTS



# GLENN ORBITS EARTH 3 TIMES SAFELY; PICKED UP IN CAPSULE BY DESTROYER; PRESIDENT WILL GREET HIM IN FLORIDA

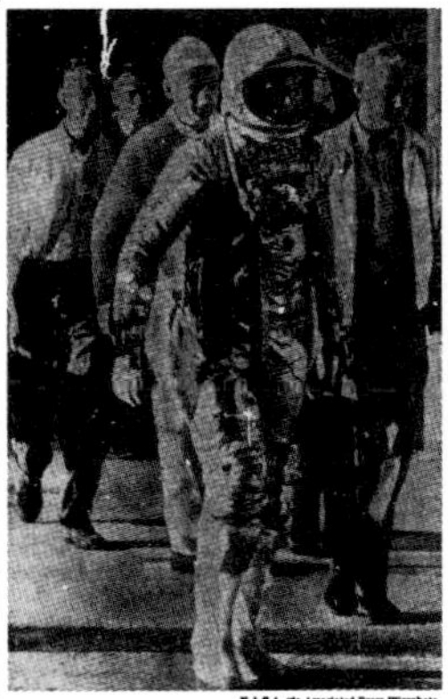

N.A.S.A. via Associated Press Wirephoto
READY: Lieut. Col. John H. Glenn Jr. walks to the van to take him to the launching site at Cape Canaveral, Fla.

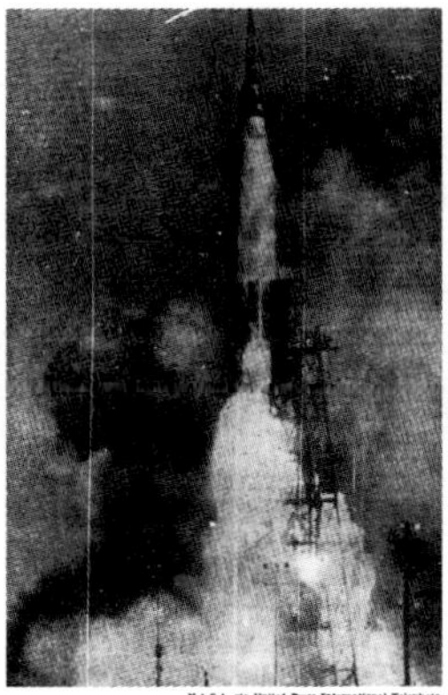

N.A.S.A. via United Press International Telephoto
LIFT-OFF: The Atlas rocket booster bearing the Project Mercury spacecraft roars aloft with 360,000-pound thrust.

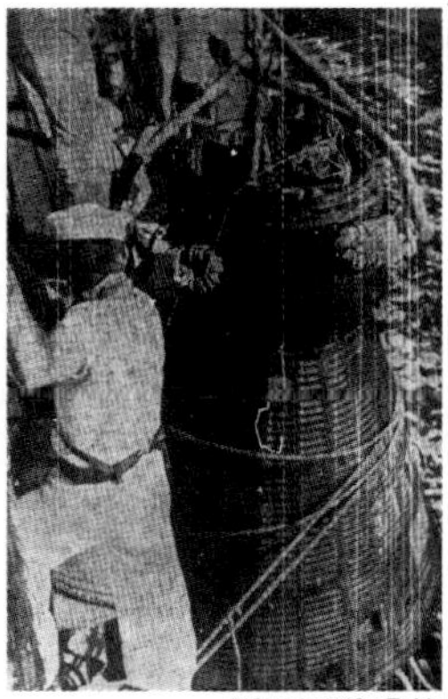

N.A.S.A. via Associated Press Wirephoto
RECOVERY: Crewmen of destroyer Noa secure capsule carrying astronaut before lifting it out of the Atlantic.

## 81,000-MILE TRIP

### Flight Aides Feared for the Capsule as It Began Its Re-Entry

*Transcript of conversations with Glenn, Pages 25 and 26.*

By RICHARD WITKIN
Special to The New York Times.

CAPE CANAVERAL, Fla., Feb. 20—John H. Glenn Jr. orbited three times around the earth today and landed safely to become the first American to make such a flight.

The 40-year-old Marine Corps lieutenant colonel traveled about 81,000 miles in 4 hours 56 minutes before splashing into the Atlantic at 2:43 P. M. Eastern Standard Time.

He had been launched from here at 9:47 A. M.

The astronaut's safe return was no less a relief than a thrill to the Project Mercury team, because there had been real concern that the Friendship 7 capsule might disintegrate as it rammed back into the atmosphere.

There had also been a serious question whether Colonel Glenn could complete three orbits as planned. But despite persistent control problems, he managed to complete the entire flight plan.

**Lands in Bahamas Area**

The astronaut's landing place was near Grand Turk Island in the Bahamas, about 700 miles southeast of here.

Still in his capsule, he was plucked from the water at 3:01 P. M. [illegible] and tackle by the destroyer Noa. The capsule was deposited on deck at 3:04.

Colonel Glenn's first words as he stepped out onto the Noa's deck were: "It was hot in here."

He quickly obtained a glass of iced tea.

He was in fine condition except for two skinned knuckles hurt in the process of blowing out the side hatch of the capsule.

The colonel was transferred by helicopter to the carrier Randolph, whose recovery helicopters had raced the Noa for the honor of making the pick-up. After a meal and extensive "de-briefing" aboard the carrier, he was flown to Grand Turk by submarine patrol plane for two days of rest and interviews on technical, medical and other aspects of his flight.

The Noa, nearest ship to the

Continued on Page 20, Column 1

## CARLINO CLEARED IN SHELTER CASE BY ETHICS PANEL

### Lane Scored in Unanimous Report, Which He Calls 'Cynical and Callous'

*Text of concluding sections of report is on Page 50.*

By WARREN WEAVER Jr.
Special to The New York Times.

ALBANY, Feb. 20—The Assembly Committee on Ethics and Guidance exonerated Speaker Joseph F. Carlino today of charges of conflict of interest made by Assemblyman Mark Lane.

In a unanimous report submitted to the Legislature, the bipartisan committee said:

¶Mr. Carlino did not "betray the public trust" by serving as a director of a company manufacturing home fall-out shelters while helping to pass school-shelter legislation last November.

¶He did not draft or support the shelter legislation "in any improper manner" for the benefit of the company, Lancer Industries, Inc.

¶The Speaker was not influenced in his official actions in behalf of the bill by the fact that he was a member of the board of directors of Lancer.

¶He did not receive any special benefit from the passage of the legislation.

**Charges Unsubstantiated**

"The committee concludes with respect to each and every accusation contained in the charges filed," the report said, "that Assemblyman Lane and those who testified in their support failed to submit credible evidence to substantiate them."

In submitting the report, the [illegible] that the full 150-member lower house vote "with respect to the conclusions reached herein" in the light of the fact that "the charges were directed against its [the Assembly's] highest ranking official."

Assemblyman Donald A. Campbell, Republican of Amsterdam, who is chairman of the committee, said he would move in the Assembly tomorrow for acceptance of the report. Mr. Carlino is expected to be absent during the debate and vote.

Assemblyman Lane, a Democrat of Manhattan, had charged that the Speaker was guilty of

Continued on Page 50, Column 1

## ROCKEFELLER BARS KOREA WAR BONUS

### Voices Opposition in Face of Legislators' Backing

By LAYHMOND ROBINSON
Special to The New York Times.

ALBANY, Feb. 20—Governor Rockefeller expressed strong opposition tonight to a state bonus for veterans of the Korean war.

Mr. Rockefeller told the New York State Department of the American Legion that he could not "as a responsible leader of government" support the demand for a bonus. The veterans' group had been campaigning for a $100,000,000 bonus for the 482,000 Korean war veterans or their next-of-kin in the state.

The Governor said his stand was backed "unanimously" by the "Republican leadership of the state." This was a reference to the leaders of the Republican-controlled Legislature.

He said that demands for funds for education, mental health, narcotics control and other state services were too great to permit a diversion of money for a veterans' bonus.

Mr. Rockefeller thus took a position in direct opposition to that of most of the Republican and Democratic members of the Legislature, who have been pushing for the bonus. The issue

Continued on Page 51, Column 1

## Jersey Bus Strike Settled; Service Is Due Tomorrow

By PETER KIHSS

An agreement to end the New Jersey bus strike was reached last night. The agreement, subject to ratification by the striking employes, was announced by Gov. Richard J. Hughes. The pact will be submitted to the union members at their garages starting at 7 A. M. today.

Union and management men expressed hope that buses could begin operating tomorrow at 4:30 A. M.

The strike against Public Service Coordinated Transport started at 12:01 A. M. Monday and halted 2,511 buses providing 1,000,000 rides a day. The company's 200 routes serve all of New Jersey's twenty-one counties except Warren and Hunterdon and go into New York City and Philadelphia. The Newark subway system was also shut.

**Carlin Gets Credit**

Governor Hughes credited Mayor Leo P. Carlin of Newark with having "sparkplugged" the successful negotiations. Mayor Carlin flew back from a Miami Beach vacation yesterday and arranged the talks with both sides and with Daniel F. Fitzpatrick, a Federal mediator, and the Governor and himself. The meeting started in Newark at 8:30 P. M., and the agreement was announced at 11:28 P. M.

Earlier, David L. Yunich, president of Bamberger's New Jersey, had asserted that the strike was having a "devastating * * * almost catastrophic" effect on retail business in Newark and elsewhere in the state. A Camden department store reported sales had fallen nearly 50 per cent on Monday, although not that far yesterday.

Despite the drop in shopping, most commuters managed to get to work by alternate means and with a minimum of confusion.

The agreement reached last night provides for a wage increase of 10 cents an hour retroactive to Feb. 1 and extending to next Feb. 1; 4 cents more an hour from then until Aug. 1, 1963, and another 4 cents an hour from then until

Continued on Page 39, Column 3

## ROSENTHAL WINS QUEENS ELECTION

### But Democrat-Liberal Has Margin of Only 193 Votes —Machines Guarded

By CLAYTON KNOWLES

Benjamin S. Rosenthal, a Democrat-Liberal backed by President Kennedy, squeaked through to victory last night in a special Congressional election in Queens' Sixth District.

By the slim margin of 193 votes, Mr. Rosenthal, a lanky 38-year-old Elmhurst lawyer, edged past Thomas F. Galvin of Flushing, the Republican candidate, to win a three-way race. Emil Levin of Flushing, a Democrat running as an independent, finished far behind.

The unofficial final tally, delayed as the early vote was hastily rechecked for errors, was: Rosenthal, 16,032; Galvin, 15,839, and Levin, 4,216.

Republicans immediately challenged the result and, while Mr. Galvin did not immediately ask for a recount, he sent a telegram demanding that the voting machines be impounded. All voting machines, normally just

Continued on Page 48, Column 3

## McNamara Reports Gains by Vietnamese

By JACK RAYMOND
Special to The New York Times.

WASHINGTON, Feb. 20—Secretary of Defense Robert S. McNamara returned to the capital today and reported improvement in the South Vietnamese effort against Communist insurgents.

He had presided at a meeting of United States military and civilian officials yesterday at the headquarters in Hawaii of Admiral Harry S. Felt, commander of United States forces in the Pacific. The meeting was the third in a series of monthly talks on the hostilities in South Vietnam.

A spokesman for Mr. McNamara said that the forces of South Vietnam, supported by the United States, "are hitting

Continued on Page 5, Column 5

## KENNEDY PRAISES 'WONDERFUL JOB'

### Tells Glenn Nation Is 'Really Proud of You'—Welcome at White House Planned

By TOM WICKER
Special to The New York Times

WASHINGTON, Feb. 20 — President Kennedy phoned Lieut. Col. John H. Glenn Jr. today immediately after the astronaut's successful orbital flight [illegible] Cape Canaveral Friday morning.

The President also set in motion plans for bringing Colonel Glenn to Washington on Monday or Tuesday, for receptions at the White House and the Capitol and a parade down Pennsylvania Avenue

A television set in his office and an open telephone line to Cape Canaveral had kept Mr. Kennedy informed of Colonel Glenn's progress all through the day.

The astronaut's three orbits around the earth, Mr. Kennedy said in a statement, have embarked the United States on a "new ocean"—that of space.

"I believe the United States must sail on it and be in a position second to none," the President said within minutes of Colonel Glenn's safe emergence from his Mercury capsule.

Colonel Glenn, he said, is the "kind of American of whom we are most proud." Mr. Kennedy also praised "all those who participated" in making the astronaut's flight successful.

Then, at 4:10 P. M., Mr. Ken-

Continued on Page 23, Column 7

## Leaders of Algeria Back Peace Terms

By THOMAS F. BRADY
Special to The New York Times.

TUNIS, Feb. 20—The Algerian nationalist Provisional Government met today and gave full approval to peace accords negotiated with the French by four members of the rebel regime.

One Algerian said afterward: "All twelve members of the Government are in unanimous agreement." This was a reference to five ministers who are prisoners in France, the four negotiators and three ministers who remained in Tunis during the secret talks last week on the French-Swiss border.

The negotiators were Belkacem Krim, M'Hammed Yazid, Saad Dahlab and Lakhdar Ben Tobbal. They met here today

Continued on Page 11, Column 1

**NEWS INDEX**

| | Page | | Page |
|---|---|---|---|
| Art | 57 | Music | 53-57 |
| Books | 43 | Obituaries | 41, 45 |
| Bridge | 42 | Real Estate | 76 |
| Business | 64, 74-75 | Screen | 53-57 |
| Buyers | 75 | Ships and Air | 90 |
| Crossword | 43 | Society | 39 |
| Editorial | 44 | Sports | 60-64 |
| Events Today | 43 | Theatres | 53-57 |
| Fashions | 38 | TV and Radio | 91 |
| Financial | 65-73 | U. N. Proceedings | 3 |
| Food | 39 | Wash. Proceedings | 29 |
| Man in the News | 29 | Weather | 99 |

News Summary and Index, Page 47

## The President's Statement

Special to The New York Times.

*WASHINGTON, Feb. 20—Following is the text of President Kennedy's statement on Colonel Glenn's flight:*

I know that I express the great happiness and thanksgiving of all of us that Colonel Glenn has completed his trip, and I know that this is particularly felt by Mrs. Glenn and his two children.

A few days ago Colonel Glenn came to the White House and visited me, and he is—as are the other astronauts—the kind of American of whom we are most proud.

Some years ago, as a Marine pilot, he raced the sun across this country—and lost. And today he won.

I also want to say a word for all those who participated with Colonel Glenn in Canaveral. They faced many disappointments and delays—the burdens upon them were great—but they kept their heads and they made a judgment, and I think their judgment has been vindicated.

We have a long way to go in this space race. We started late. But this is the new ocean, and I believe the United States must sail on it and be in a position second to none.

Some months ago I said that I hoped every American would serve his country. Today Colonel Glenn served his, and we all express our thanks to him.

## ADENAUER WANTS PARLEY ON BERLIN

### Suggests Foreign Ministers of Big Four Meet 'Soon'

By SYDNEY GRUSON
Special to The New York Times.

BONN, Germany, Feb. 20—Chancellor Adenauer suggested today that a Big Four foreign ministers' conference on Berlin should be convened "soon." He was speaking to the Parliamentary group of the Christian Democratic Union.

He said that it might be "expedient" to "take a pause" in the Berlin talks now going on between Andrei A. Gromyko, the Soviet Foreign Minister, and Llewellyn E. Thompson Jr., the United States Ambassador to Moscow.

Ambassador Thompson should not continue "negotiating" endlessly, Dr. Adenauer added. There have been four meetings in the last seven weeks between Mr. Gromyko and Mr. Thompson without any advance toward a Berlin settlement.

[A warning by Izvestia, the Soviet Government newspaper, that Moscow was ready to push through a separate peace treaty with East Germany if the United States did not alter its position [illegible] the talks raised the possibility of a renewal of the Soviet deadline on a peace pact.]

Dr. Adenauer's advocacy of a new conference of the United States, British, French and Soviet foreign ministers reflected his unhappiness with the course of the Gromyko-Thompson talks.

He is known to believe that Mr. Thompson has made what

Continued on Page 2, Column 2



## URBAN PLAN VOTE PUT OFF IN SENATE

### Administration Rebuffed on Forcing Issue to Floor

By RUSSELL BAKER
Special to The New York Times.

WASHINGTON, Feb. 20 — President Kennedy affronted the Senate's dignity today and got a political rebuff for it.

In a surprising repudiation of the Administration's voting form sheets, the elders turned on the White House and rejected a leadership move to get a quick floor test of the President's urban affairs proposal. The vote was 58 to 42.

Thus, the White House lost its chance to get a favorable Senate vote on the plan before the House could vote to kill it. The Democrats also lost their chance to get the Senate's Republicans clearly on record for or against the plan to create a Cabinet-level Department of Urban Affairs and Housing.

Today's test came on the dusty parliamentary question whether the Senate should take the plan away from the Government Operations Committee and bring it to an immediate floor vote. This is known as "discharging" the committee. It is an extraordinary procedure that is rarely used because it is repugnant to Senate traditions.

Today it became the instrument of the President's defeat.

The move to discharge the Government Operations Committee was undertaken with misgivings yesterday by Mike Mansfield of Montana, Senate Democratic leader. The reason was a sudden threat by the

Continued on Page 16, Column 4

## COL. GLENN FLOWN TO ISLE FOR CHECK

### He Feels Tired but Elated —Goes to Grand Turk for Report and Examination

By JOHN W. FINNEY
Special to The New York Times

GRAND TURK ISLAND, Feb. 20—An elated but tired John H. Glenn Jr. returned to earth tonight and reported that he "couldn't feel better."

[illegible] also reported that he had felt no sickness or discomfort during his five-hour, three-orbit flight around the earth, even during the extended period of weightlessness.

Colonel Glenn landed at this small British possession at 9:11 P. M. in a Navy S-2-F submarine patrol plane. He was clad in light blue coveralls. He had co-piloted the plane from the carrier Randolph, where he spent several hours after being retrieved from the Atlantic ocean.

Around his ears were the marks of the earphones that he had worn while piloting a plane that traveled at about one-hundredth the speed of his Friendship 7 space capsule. And on his face was an excited enthusiastic smile.

Asked how he felt, the red-headed marine replied: "Fine, wonderful, I couldn't feel better."

And he was also hungry. His first comment on stepping into the small hospital arranged for him was: "First I want something to eat—I am hungry." A steak dinner was promptly or-

Continued on Page 23, Column 2

## Moscow, Unmoved, Gives News of Orbit

By THEODORE SHABAD
Special to The New York Times.

MOSCOW, Feb. 20 — The Russians voiced congratulations tonight on hearing of Lieut. Col. John H. Glenn Jr.'s orbital space flight.

But they showed no enthusiasm on the successful launching and landing of the spacecraft Friendship 7.

These reactions were reported from Moscow University by United States exchange students who had been listening with Russians to radio reports of Colonel Glenn's progress.

"They congratulated us in friendly fashion but were oddly reserved," an American said.

Soviet radio and television were unusually prompt in reporting the flight. The first bul-

Continued on Page 22, Column 5

## NEW YORK PAUSES TO 'WATCH' GLENN

### Millions Rivet Attention on Astronaut in Flight

By NAN ROBERTSON

The thoughts of millions of New Yorkers were riveted for hours yesterday on one man alone in space.

Minute by minute, they followed the orbital flight of Lieut. Col. John H. Glenn Jr. three times around the earth, waiting in agonizing suspense for his safe return. The life of New York almost stood still during the dramatic countdown.

From then on until Colonel Glenn scrambled "hale and hearty" out of his capsule on the destroyer Noa, people carried on absent-mindedly and in spurts. Millions of working hours were lost during the day, but no one could have begrudged this. Employers and the employed alike were drawn irresistibly to radio and television sets.

The most spectacular display of interest occurred in Grand Central Terminal, where throngs of up to 9,000 persons massed before a huge television screen. The police described it as the largest static crowd in the station's history. The terminal manager said those who

Continued on Page 22, Column 6

Glenn found the feeling of weightlessness to be very pleasant. He had no trouble eating food from squeeze tubes. He said the view was "tremendous."

Glenn's flight, which lasted four hours and fifty-five minutes, was a great success. He orbited the earth three times.

When it came time to fire the retrorockets to slow down the capsule for reentry, Glenn had some exciting moments. All three rockets went off as scheduled. But as the capsule reentered the atmosphere, a retro-pack flew off in "big, flaming chunks." Glenn heard "small things brushing against the capsule." He said he saw "a real fireball outside."

But Glenn brought the capsule through without any mishap. The main parachute popped open on schedule and *Friendship 7* splashed down to a comfortable landing in the Atlantic Ocean.

Glenn was an instant hero. Four million people cheered and applauded him at a parade in his honor in New York City on John Glenn Day. Glenn resigned from the space program in 1964. Ten years later, he won election to the United States Senate. He was reelected in 1980 and 1986.

In the years that followed Glenn's flight, more and more American astronauts took their turns orbiting the earth. Unmanned spacecraft explored the moon and brought back soil samples to Earth. And plans were drawn for the first manned flight to the moon.

*John Glenn as photographed by an automatic camera inside his* Mercury *spacecraft. A weightless tube of applesauce floats to his left.*

*John Glenn, the first American to orbit the earth.*

# Famous Firsts

*Valentina Tereshkova, the first woman in space, undergoes a check of her pulse rate*

In the years after the orbital flights of John Glenn and Yuri Gagarin, one breakthrough in space followed another. They included:

### *First Woman in Space June 16, 1963*

After Yuri Gagarin's orbital flight in 1961, officials of the Soviet Union's space program received hundreds of letters from people who wanted to become astronauts.

Many of the letters were from women. From those who wrote, Valentina Tereshkova, a 24-year-old amateur parachutist who worked in a textile factory, was chosen by Soviet officials as the woman who should represent women of the world in space.

After eighteen months of training, Tereshkova was launched into orbital flight aboard *Vostok 6*. The mission created a sensation. Tereshkova was pictured floating weightlessly in her space capsule. She sent "best wishes" to the American people and "warm greetings" to the Chinese.

Three days and 48 orbits after takeoff, Tereshkova's capsule reentered the earth's atmosphere. She was ejected from her capsule and floated to Earth by parachute, landing near the small village of Karaganda in the republic of Kazakhstan.

The empty *Vostok 6*, slowed by another parachute, landed nearby.

In a telephone conversation with Soviet leader Nikita Khrushchev, Tereshkova admitted she had bruised her nose when she landed. Three days later, when she arrived in Moscow for her official reception, Khrushchev greeted her with a hug and a kiss.

*Soviet citizens in Moscow celebrate the news of Tereshkova's safe landing.*

In those days, women were not accepted in many fields. The U.S. space program, for example, involved no women astronauts at the time. Tereshkova's flight was an inspiration to women in every part of the world.

## Launching of the First Communications Satellite April 6, 1965

In 1964, the United States joined eleven other nations to form an organization called INTELSAT, for International Telecommunication Satellite Consortium. INTELSAT's purpose was to build and launch satellites that would link together the telephone and television networks of many countries.

The first of these satellites was launched from Cape Kennedy on April 6, 1965. It was called *Early Bird* at first, but later it was named *Intelsat 1*.

Weighing 85 pounds and cylinder-shaped, *Intelsat 1* was guided into orbit 22,300 miles above the Atlantic Ocean.

It was thus able to provide communications service between the United States and Europe.

The satellite appeared to be stationary in the sky, but it wasn't. It completed an orbit every twenty-four hours, the same amount of time it takes the earth to make a full turn. In other words, the satellite "kept up" with the earth. It was as though it was at the end of a 22,300-mile-long pole that extended upward from the earth.

*Intelsat 1* handled 240 telephone calls at one time, or one television channel.

By 1990, more than thirty Intelsat satellites had been put in orbit. Each new satellite is more advanced than the one before it. One of the latest models handles 33,000 telephone calls at one time and is bigger than a passenger automobile. But it was *Intelsat 1* that paved the way.

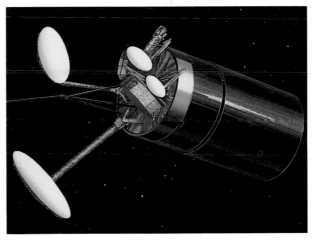

*This is* Intelsat 6, *a communications satellite that was scheduled to be launched in 1989. Nearly 39 feet high, it's the biggest communication satellite ever put into space.*

## First American to Walk in Space June 3, 1965

When astronauts Edward White and James McDivitt, both Air Force officers, lifted off from Cape Kennedy in a *Gemini 4* spacecraft, they had a special mission.

*Secured to his* Gemini *spacecraft by a 23-foot-long tether line, Edward White performs his space walk.*

It was officially known as "extravehicular activity." But everyone called it a space walk.

On their third orbit, while the spacecraft was traveling 120 miles above the Pacific Ocean at 17,500 miles an hour, White, after struggling with a balky hatch, popped out of the capsule. McDivitt stayed inside taking pictures of him.

As he floated in space, White remained secured to the spacecraft by a 23-foot tether line. An air hose, which was taped to the tether line, supplied his oxygen.

White tumbled about by means of a zip gun, which squirted out bursts of compressed air. After about fifteen minutes,

the gun ran out of fuel, and Mission Control ordered White back inside. He didn't want to go; he was having too much fun. "It's the saddest moment of my life," White radioed.

Three months before, Aleksei Leonov, a Soviet astronaut had drifted in space from the end of a cord attached to his spacecraft. He was the first person to walk in space.

White was second. But the fact that he was not first did not dampen White's enthusiasm at the idea of being able to open the capsule's hatch and step out into space while traveling at almost 18,000 miles an hour.

## First Manned Orbit of the Moon December 24, 1968

By the late 1960s, American scientists and engineers had decided how they would put a three-person crew of astronauts on the moon.

A Saturn 5 rocket, launched from Earth, would boost an *Apollo* spacecraft into orbit around the moon. But the entire spacecraft would not land on the moon. Instead, a small landing craft — called a lunar module — would separate from the spacecraft and set down on the lunar surface. One of the astronauts would remain on board the spacecraft, keeping it in orbit around the moon, while the other two explored and conducted scientific experiments.

When their work was done, the two astronauts would blast off in the landing craft and rejoin the spacecraft. The three astronauts would then rocket back to Earth.

*Apollo 8* was a test in which the cone-shaped spacecraft was to be put into orbit around the moon.

Crew members aboard *Apollo 8* were William Anders, James Lovell, Jr., and Frank Borman.

Just eleven minutes after lift-off, *Apollo 8* went into orbit around the earth. During its second orbit, the Saturn rocket fired again, sending the spacecraft toward the moon at a speed of 25,000 miles an hour.

The crew of *Apollo 8* spent Christmas Eve circling the moon. They marked the event by reading from the Bible and concluding their TV broadcast with, ". . . Merry Christmas. God bless all of you — all of you on the good earth." On Christmas morning, the *Apollo 8* crew fired the rocket that boosted them back to Earth.

From beginning to end, it was a near-perfect performance. In its 147-hour flight, the spacecraft traveled more than half a million miles, including 10 orbits of the moon. A moon landing was to be the next step.

*Photograph of the nearly full moon taken from the* Apollo 8 *spacecraft.*

# "The *Eagle* Has Landed"

They called *Apollo 11* the "big one." It was not just another space flight, not just another test. *Apollo 11* was going to put humans on the moon.

The flight of the "big one" began right on schedule. On July 16, 1969, at 9:32 A.M., bright orange flames and dark smoke began pouring out of the Saturn 5 rocket that supported the *Apollo* spacecraft. The powerful blast-off battered eardrums and made the ground tremble.

Neil Armstrong, Edwin Aldrin, Jr., and Michael Collins were forced back into their seats as the giant rocket lifted into the sky. Within two and a half minutes, they were streaking upward at 6,200 miles an hour.

After $1\frac{1}{2}$ orbits of the earth, another stage of the Saturn 5 fired, accelerating the spacecraft to a speed of 24,245 miles an hour, just enough to tear it from the earth's gravitational pull and send it toward the moon.

In their four-day journey, the astronauts saw little that was new. They followed a trail that had been blazed by *Apollo 8* and other *Apollo* missions.

During this time, the astronauts checked equipment and systems, carried out household tasks, and conducted telecasts for the people back home. "We have a happy home," Collins said during one of the telecasts. "Plenty of room for the three of us."

*A plume of flame signals the liftoff of the* Apollo 11 *space vehicle and the mission that put American astronauts on the moon.*

*A view of the rising earth taken from the* Apollo *spacecraft. The lunar horizon is in the foreground.*

A tense moment came when the astronauts neared the moon and then went behind it. For thirty-three minutes, *Apollo 11* was out of radio contact with the earth.

During this time, the astronauts had to fire a rocket to slow down the vehicle so it would go into lunar orbit. If the rocket failed to fire, the craft would simply loop the moon and head back to Earth. If the rocket fired for too long, they would crash-land on the moon. But the rocket performed without a hitch, and the spacecraft swept into moon orbit.

After twenty-four hours in orbit, Armstrong and Aldrin climbed from the mother ship, which they had named *Columbia*, into the lunar landing craft, *Eagle*, which was linked to it. They then began preparing for their descent to the lunar surface. Collins, meanwhile, continued to pilot *Columbia* in lunar orbit.

On their thirteenth orbit of the moon, Armstrong fired the rockets that gently separated *Eagle* from *Columbia*. Then Armstrong slowed the lunar module's engine so *Eagle* could drop to the moon.

On July 20, the four-legged, spidery landing craft settled down on the moon's Sea of Tranquility, an open plain. Armstrong radioed to Earth, "Houston, Tranquility Base here. The *Eagle* has landed."

*Astronaut Edwin Aldrin descends the steps of the lunar module* Eagle *to the surface of the moon.*

Loud cheers and frantic applause erupted at Mission Control in Houston.

There were more tense moments on July 21 when Armstrong and Aldrin were preparing to leave the moon and had to fire the ascent rocket in the upper portion of the lunar module. It *had* to fire; there was no backup. It worked perfectly, lifting *Eagle* toward a hookup with Collins and *Columbia*.

Besides the lower portion of the lunar module, the astronauts left behind other debris, including their backpacks, boots, and other equipment they no longer needed. They also left an American flag, a plaque with President Nixon's signature, and microfilmed messages from the leaders of 72 foreign nations — and the footprints of the first humans to walk on the moon.

In the years that followed, American astronauts made several other visits to the moon. There were, in fact, six moon landings between 1969 and 1972.

*Apollo 17*, in December, 1972, was the last moon mission. The astronauts of *Apollo 17* were Eugene Cernan, Ronald Evans, and Dr. Harrison Schmitt, a geologist, and the first scientist to be assigned to a space mission. They left behind a plaque that read:

Here man completed his first
Explorations of the moon
December 1972, A.D.
May the spirit of peace in which we came
Be reflected in the lives of all mankind.

On the Apollo 17 mission to the moon, the crew used a small four wheeled vehicle (seen at right) to explore the lunar surface. Here scientist-astronaut Harrison Schmitt examines a lunar boulder.

Aldrin walking near one of the scientific experiments the astronauts set up on the moon's surface. Lunar landing craft Eagle is at right.

# Living in Space

*A close-up of the* Skylab *space station photographed against a black sky. The large rectangles are solar panels.*

Beginning in the early 1970s, the United States and the Soviet Union began to experiment with manned space stations that orbited the earth. A space station is a special kind of satellite, one in which astronauts can live and work for weeks or even months at a time.

*Skylab*, the first United States space station, was put into orbit early in 1973. Nine astronauts in three crews spent twenty-eight, fifty-nine, and eighty-four days in Earth orbit aboard *Skylab*.

The first crew arrived in May, 1973, docking alongside *Skylab* in an Apollo-type vehicle.

Their "house in space" consisted of a huge metal cylinder nearly 50 feet in length and 20 feet in diameter. The cylinder was divided into two floors.

The upper floor, used mostly for storage, was ringed with lockers. The lower floor had three tiny bedrooms, a bathroom, and an all-purpose room that could be used as a living room, dining room, and work area.

Each bedroom was a tiny compart-

*Astronaut Gerald Carr demonstrates weightlessness during* Skylab 4 *mission.*

ment, just big enough to hold a sleeping bag. "Weightlessness is very comfortable for sleeping," said astronaut Michael Collins in his book, *Liftoff*. It's "just floating," with the light touch of the sleeping bag helping to prevent one from "straying off somewhere."

It's difficult to keep fit in space. On Earth, when you walk up stairs or run, your heart and lungs have to work hard to overcome gravity's pull. Gravity also helps to keep your leg muscles firm, since they support your upper body.

In weightlessness, the heart and lungs grow weaker since they don't have to work hard. Leg muscles begin to waste away.

Aboard *Skylab*, to keep fit, astronauts had to spend part of each day working on the stationary bicycle and treadmill with which the space station was equipped.

The astronauts aboard *Skylab* found that their bodies stretched out a bit

without gravity pressing down on them. Some found they became as much as two inches taller aboard *Skylab* than they had been on Earth.

With three crews of three men, each occupying *Skylab* for many weeks at a time, food for over a thousand meals had to be stored aboard the spacecraft. Equipment for conducting several different types of scientific experiments also had to be carried.

There were, in fact, more than 12,000 items stored aboard *Skylab*.

On Earth, gravity helps keep things in place. But open a storage compartment in space, and dozens of objects can suddenly tumble out and float away. Astronauts called this the jack-in-the box effect.

While the astronauts found their home in space to be comfortable enough, they complained about the food. It didn't taste as good as it had when they sampled it on the ground. The first *Skylab* crew

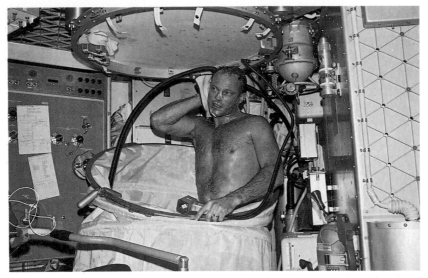

Astronaut Jack Lousma takes a shower in the crew's quarters during the flight of Skylab 3. Water, which comes from a push-button shower head attached to a flexible hose, must be drawn off by a vacuum.

Astronaut Charles Conrad, Jr., trims the hair of pilot Paul Weitz during their 28-day Skylab 2 mission. Weitz holds a vacuum hose in his right hand to collect clippings.

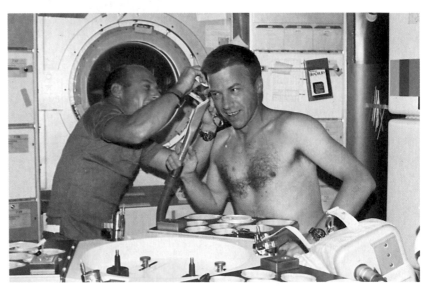

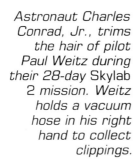

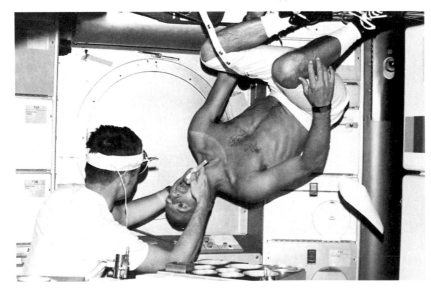

Scientist-astronaut Joseph Kerwin gives an oral examination to Charles Conrad aboard Skylab 2. Because of weightlessness in space, Conrad almost stands on his head.

suggested that later crews carry sauces and seasonings to brighten up their meals.

The *Skylab* crews conducted many scientific experiments. They used a solar telescope to study the earth and its natural resources. They also studied the sun. But the chief goal of the *Skylab* crews was to study the effects on humans of long periods spent in weightlessness.

*Skylab* stayed in space for six years. During that time, it circled the earth 35,000 times, traveling nearly one billion miles.

In 1979, on its final orbit, *Skylab* caused a brief meteor shower as it broke up upon entering the earth's atmosphere. A 17-year-old truck driver from the town of Esperance in southwestern Australia claimed a $10,000 prize from a San Francisco newspaper for recovering a chunk of America's first space station.

*On the* Skylab 3 *mission, scientist-astronaut Owen Garriott works at the Apollo Telescope Mount of the* Skylab *space station.*

# Touchdown, *Columbia*

The time: April 14, 1981.

The place: A dry lake bed in the Mojave Desert of Southern California.

The event: The return to Earth of the space shuttle *Columbia*. For the first time in history, a constructed machine was coming back from space to land like an ordinary airplane.

*Columbia* was born when scientists and engineers began trying to figure out how to deliver supplies and equipment to an operating space station, such as *Skylab*. A new spacecraft for each delivery would cost too much. The solution was a reusable vehicle, one that could fly back and forth between the earth and the space station over and over again.

The vehicle is officially known as the Space Transportation System. But everyone calls it the shuttle.

The stubby-winged shuttle, a combination rocket ship—plane, is like no other vehicle. It has three main stages:

- An orbiter: an airplane-like body with delta-shaped wings and three main engines.
- An external tank, which feeds propellants (Liquid hydrogen and oxygen) to the orbiter's engines during the first stages of flight.
- Two solid rocket boosters. These, along with the Orbiter's main engines, provide the energy for lift-off.

At lift-off, the orbiter rides piggyback on the external tank. The two solid rocket

*At the end of the first shuttle mission, orbiter* Columbia *touches down on a runway at Edwards Air Force Base in California.*

*The launch of the first space shuttle at Kennedy Space Center on April 12, 1981.*

boosters are positioned alongside the tank.

About two minutes after lift-off, the two boosters separate from the orbiter and return to Earth by parachutes. The boosters can be used again on another flight.

Just before going into orbit, the orbiter releases the external tank. The tank, the only part of the system that is not recovered and reused, breaks up over the ocean.

After a mission has been completed, the orbiter enters the earth's atmosphere, then lands on a runway like an airplane.

By the late 1980s, NASA had developed a fleet of four shuttles. Named after famous sailing ships, they were called:

*Columbia*, named after a Navy frigate launched in 1836, one of the first American sailing ships to sail around the world.

*Challenger*, named in honor of a Navy vessel that explored the Atlantic and Pacific Oceans from 1872 to 1876.

*Discovery*, named after two ships: Henry Hudson's, which in 1610 sailed in search of a northwest passage between the Atlantic and Pacific Oceans; and Captain James Cook's, a vessel that discovered the Hawaiian Islands in 1778 and explored southern Alaska and western Canada.

*Atlantis*, named after a sailing ship operated by the Woods Hole Oceanographic Institute from 1930 to 1936.

Each of the shuttles was designed to undertake at least one hundred missions. Each will travel some 90 million miles before being taken out of service.

On that April afternoon in 1981, millions of television viewers in the United States and perhaps hundreds of millions around the world watched the final stages of *Columbia*'s first venture into space. John Young, the 50-year-old commander of the flight, was at the controls.

*Columbia* functioned as a 102-ton glider; there was no engine to correct its course.

VOYAGE OF COLUMBIA: A SPECIAL PULLOUT SECTION BEGINS ON PAGE A9

"All the News That's Fit to Print"

The New York Times

LATE CITY EDITION

Weather: Mostly cloudy, cool today; periods of rain tonight and tomorrow. Temperature range: today 42-52; yesterday 49-68. Details are on page B6.

VOL.CXXX . . . No. 44,917 | Copyright © 1981 The New York Times | NEW YORK, MONDAY, APRIL 13, 1981 | 30 cents beyond 50-mile zone from New York City. Higher in air delivery cities. | 25 CENTS

# SHUTTLE ROCKETS INTO ORBIT ON FIRST FLIGHT; SOME TILES FALL OFF, BUT NASA SEES NO DANGER

## POLISH COMMUNISTS ASSURE THEIR ALLIES THEY WILL END CRISIS

### Criticized at East German Meeting for Not Taking Strong Action, Warsaw Asks for Time

Special to The New York Times

EAST BERLIN, April 12 — The Polish Communists, facing mounting pressure from their allies for speedy action, pledged here today to seek a "political solution" of the Polish crisis by mobilizing the party for a struggle against enemies of Communism and "counterrevolutionaries."

Speaking at the East German party congress in the presence of Warsaw Pact allies and delegates from Communist parties all over the world, Kazimierz Barcikowski, a member of the Polish Politburo, said the party was determined to assert its authority and "find a way to settle the complicated and difficult problems that have arisen in Poland as a result of the severe social and economic crisis."

Mr. Barcikowski's indirect request for more time to resolve Poland's difficulties came after rank-and-file East German delegates had bluntly urged a crackdown on dissidents, saying party action was "overdue."

**Indirect Message to Poles**

In his opening remarks yesterday, the East German party leader, Erich Honecker, voiced guarded support for the Polish party's ability to overcome its problems, but by letting low-level aides express distaste and impatience at the developments in the neighboring country, he let the Poles know time may be running out.

Mikhail A. Suslov, the veteran Soviet theoretician and head of the Soviet dele-

Continued on Page A4, Column 3

## Rioting Sweeps London District For Second Day

### Dozens of Police Injured in Clashes With Blacks

By WILLIAM BORDERS
Special to The New York Times

LONDON, April 12 — Hundreds of youths, most of them black, rampaged through the Brixton section of south London this evening in a second night of clashes with the police.

Hurling stones, bottles and homemade firebombs at the advancing ranks of patrolmen, the rioters set fire to several shops and overturned more than a dozen automobiles, setting most afire.

But with more than 1,000 policemen on emergency duty in an area of no more than 20 blocks, tonight's disturbances were far less severe than the ones last night in the run-down neighborhood, which is two miles south of central London.

**Dozens of Policemen Injured**

In the two nights of disorders, more than 30 policemen have suffered injuries requiring hospitalization and dozens of others have been less seriously hurt. At least 20 civilians have been injured, two dozen buildings have been destroyed, and there have been nearly 200 arrests.

"It's the worst thing I've seen since the war," said an elderly shopkeeper in the area, surveying a block in which two buildings were still smoldering this morning. "We are not accustomed to this kind of violence, this horrible, wanton destruction."

Most of the fury of the mobs seemed to be directed at the police, who spent long hours last night and again tonight crouched behind clear plastic riot shields. Although virtually all the policemen are white, and most of the rioters were black, community leaders insisted that race was not the only cause of the

Continued on Page A17, Column 1

United Press International

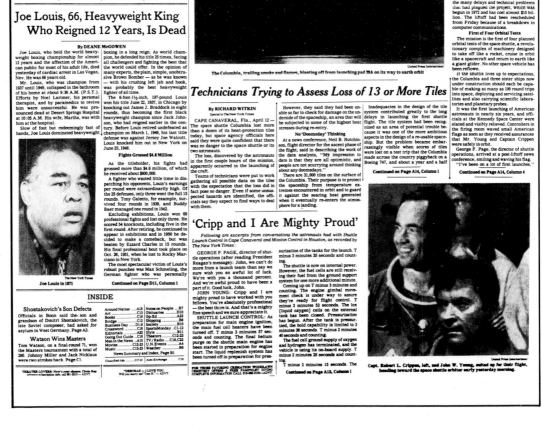

The Columbia, trailing smoke and flames, blasting off from launching pad 39A on its way to earth orbit

## LIFTOFF ON SCHEDULE

### Crippen and Young Finish Several Crucial Tasks Before First Meal

By JOHN NOBLE WILFORD
Special to The New York Times

CAPE CANAVERAL, Fla., April 12 — The space shuttle Columbia, its rockets spewing orange fire and a long trail of white vapor, blasted its way into earth orbit today, carrying two American astronauts on a daring journey to test the world's first re-usable spaceship.

Soon after they settled into orbit, John W. Young, a civilian, and Capt. Robert L. Crippen of the Navy focused a television camera on the Columbia's tail section and discovered that more than a dozen heat-shielding tiles had ripped off, possibly because of the stresses of launching.

Project officials said that the tile loss should not shorten the flight or endanger the lives of the astronauts when the Columbia plunges back into the atmosphere, glowing red-hot from frictional heat, to attempt a runway landing Tuesday. The projected 36-orbit, 54½-hour flight is scheduled to end at Edwards Air Force Base in California.

**'We've Got a Super Vehicle'**

"I'm just not concerned about it," Neil B. Hutchinson, a flight director at Mission Control in Houston, said in discussing the tile problem at a news conference this afternoon. "We've got a super vehicle up there."

Mr. Hutchinson said that the small gaps in the fragile silica-fiber tile coating were in a "noncritical area," the two identical pods housing Columbia's orbital maneuvering rockets, and that an underlayer of insulation "appears to be intact."

"It's not going to bother us on the way home," Mr. Hutchinson said.

The launching of the Columbia occurred on time at 7 A.M., after a smooth countdown that was in notable contrast to the many delays and technical problems that had plagued the project, which was begun in 1972 and has cost almost $10 billion. The liftoff had been rescheduled from Friday because of a breakdown in computer communications.

**First of Four Orbital Tests**

The mission is the first of four planned orbital tests of the space shuttle, a revolutionary complex of machinery designed to take off like a rocket, cruise in orbit like a spacecraft and return to earth like a giant glider. No other space vehicle has been reflown.

If the shuttle lives up to expectations, the Columbia and three sister ships now under construction should each be capable of making as many as 100 round trips into space, deploying and servicing satellites and also carrying scientific laboratories and planetary probes.

It was the first launching of American astronauts in nearly six years, and officials at the Kennedy Space Center were elated and visibly relieved. Controllers in the firing room waved small American flags as soon as they received assurances that Mr. Young and Captain Crippen were safely in orbit.

George F. Page, the director of shuttle operations, arrived at a post-liftoff news conference, smiling and waving his flag.

"I've been on a lot of first launches,"

Continued on Page A14, Column 4

## Joe Louis, 66, Heavyweight King Who Reigned 12 Years, Is Dead

By DEANE McGOWEN

Joe Louis, who held the world heavyweight boxing championship for almost 12 years and the affection of the American public for most of his adult life, died yesterday of cardiac arrest in Las Vegas, Nev. He was 66 years old.

Mr. Louis, who was champion from 1937 until 1949, collapsed in the bathroom of his home at about 9:30 A.M. (P.S.T.). Efforts by Noel Larimer, his personal therapist, and by paramedics to revive him were unsuccessful. He was pronounced dead at Desert Springs Hospital at 10:05 A.M. His wife, Martha, was with him at the hospital.

Slow of foot but redeemingly fast of hands, Joe Louis dominated heavyweight boxing in a long reign. As world champion, he defended his title 25 times, facing all challengers and fighting the best that the world could offer. In the opinion of many experts, the plain, simple, unobtrusive Brown Bomber — as he was known — with his crushing left jab and hook, was probably the best heavyweight fighter of all time.

The New York Times

Joe Louis in 1971

The 6-foot-1½-inch, 197-pound Louis won his title June 22, 1937, in Chicago by knocking out James J. Braddock in eight rounds, thus becoming the first black heavyweight champion since Jack Johnson, who had reigned earlier in the century. Before Louis retired undefeated as champion on March 1, 1949, his last title defense was against Jersey Joe Walcott. Louis knocked him out in New York on June 25, 1948.

**Fights Grossed $4.6 Million**

As the titleholder, his fights had grossed more than $4.6 million, of which he received about $800,000.

A fighter who wasted little time in dispatching his opponents, Louis's earnings per round were extraordinarily high. Of the 25 defenses, only three went the full 15 rounds. Tony Galento, for example, survived four rounds in 1939, and Buddy Baer managed one round in 1942.

Excluding exhibitions, Louis won 68 professional fights and lost only three. He scored 54 knockouts, including five in the first round. After retiring, he continued to appear in exhibitions and in 1950 he decided to make a comeback, but was beaten by Ezzard Charles in 15 rounds. His final professional bout took place on Oct. 26, 1951, when he lost to Rocky Marciano in New York.

The most spectacular victim of Louis's robust punches was Max Schmeling, the German fighter who was personally

Continued on Page D11, Column 1

## Technicians Trying to Assess Loss of 13 or More Tiles

By RICHARD WITKIN
Special to The New York Times

CAPE CANAVERAL, Fla., April 12 — The space shuttle Columbia lost more than a dozen of its heat-protection tiles today, but space agency officials here said they were quite confident that there was no danger to the space shuttle or its two astronauts.

The loss, discovered by the astronauts in the first couple hours of the mission, apparently occurred in the launching of the craft.

Teams of technicians were put to work gathering all possible data on the tiles with the expectation that the loss did in fact pose no danger. Even if some unsuspected hazards are identified, the officials say they expect to find ways to deal with them.

However, they said they had been unable so far to check for damage on the underside of the spaceship, an area that will be subjected to some of the highest heat stresses during re-entry.

**No 'Doomsday' Thinking**

At a news conference, Neil B. Hutchinson, flight director for the ascent phase of the flight, said in describing the work of the data analysts, "My impression to date is that they are all optimistic, and people are not scurrying around thinking about any doomsdays."

There are 31,000 tiles on the surface of the Columbia. Their purpose is to protect the spaceship from temperature extremes encountered in orbit and to guard it against the searing heat generated when it eventually re-enters the atmosphere for a landing.

Inadequacies in the design of the tile system contributed greatly to the long delays in launching the first shuttle flight. The tile system had been recognized as an area of potential trouble because it was one of the more ambitious aspects in the design of a re-usable spaceship. But the problem became embarrassingly visible when scores of tiles were lost on a test trip that the Columbia made across the country piggyback on a Boeing 747, and about a year and a half

Continued on Page A14, Column 1

## 'Cripp and I Are Mighty Proud'

*Following are excerpts from conversations the astronauts had with Shuttle Launch Control in Cape Canaveral and Mission Control in Houston, as recorded by The New York Times:*

GEORGE F. PAGE, director of shuttle operations (after reading President Reagan's message): John, we can't do more from a launch team than say we sure wish you an awful lot of luck. We're with you a thousand percent. And we're awful proud to have been a part of it. Good luck, John.

JOHN YOUNG: Cripp and I are mighty proud to have worked with you fellows. You're absolutely professional — the best there is. And that's a mighty fine speech and we sure appreciate it.

SHUTTLE LAUNCH CONTROL: As preparation for main engine ignition, the main fuel cell heaters have been turned off. T minus 3 minutes 57 seconds and counting. The final helium purge on the shuttle main engine has been started in preparation for engine start. The liquid replenish system has been turned off in preparation for pressurization of the tanks for the launch. T minus 3 minutes 35 seconds and counting.

The shuttle is now on internal power. However, the fuel cells are still receiving their fuel from the ground support system for one more additional minute.

Coming up on T minus 3 minutes and counting. The engine gimbal movement check is under way to assure they're ready for flight control. T minus 2 minutes 52 seconds. The lox [liquid oxygen] cells on the external tank has been closed. Pressurization has begun. After the tank is pressurized, the hold capability is limited to 3 minutes 38 seconds. T minus 2 minutes 40 seconds and counting.

The fuel cell ground supply of oxygen and hydrogen has terminated, and the vehicle is using its on-board supply. T minus 2 minutes 25 seconds and counting.

T minus 2 minutes 15 seconds. The

Continued on Page A13, Column 1

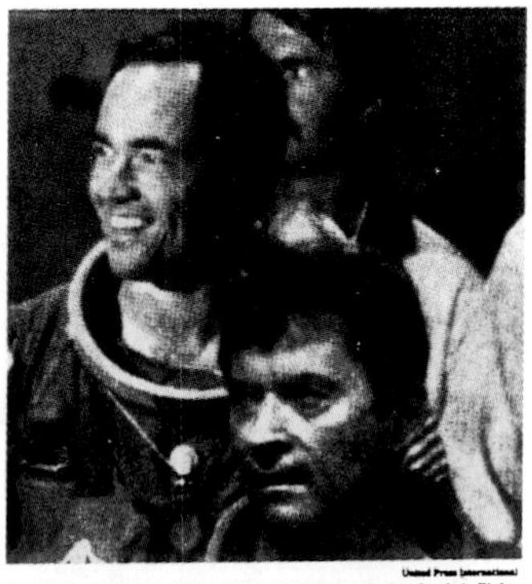

United Press International

Capt. Robert L. Crippen, left, and John W. Young, suited up for their flight, heading toward the space shuttle orbiter early yesterday morning.

## INSIDE

**Shostakovich's Son Defects**

Officials in Bonn said the son and grandson of Dmitri Shostakovich, the late Soviet composer, had asked for asylum in West Germany. Page A3.

**Watson Wins Masters**

Tom Watson, on a final-round 71, won the Masters tournament with a total of 280. Johnny Miller and Jack Nicklaus were two strokes back. Page C1.


| | | | |
|---|---|---|---|
| Around Nation | A18 | Notes on People | B7 |
| Art | C13 | Obituaries | D10 |
| Books | C16 | Op-Ed | A23 |
| Bridge | C12 | Shipping | B8 |
| Business Day | D1-9 | Society | B10 |
| Crossword | C16 | SportsMonday | C1-12 |
| Editorials | A22 | Style | B11 |
| Going Out Guide | C16 | Theaters | C12-23 |
| Men in the News | A10 | TV/Radio | C16,C22 |
| Movies | C12-23 | U.N. Events | A4 |
| Music | C12-23 | Weather | B6 |

News Summary and Index, Page B1

| | | | |
|---|---|---|---|
| Classified Ads | C17-21 | Auto Exchange | C18 |

*Astronaut John Young, commander of the first shuttle flight, on board the orbiter* Columbia, *prepares to make a logbook entry.*

*Astronaut Robert Crippen, who piloted* Columbia *on the first shuttle mission, takes advantage of weightlessness to try some acrobatics.* Columbia *was in earth orbit at the time.*

Nevertheless, Young lined up *Columbia* perfectly with the runway that had been marked on the flat, dry lake bed.

Nineteen seconds before landing, Young dropped the wheels. "Gear down," reported the pilot of a chase jet.

Then *Columbia*'s rear wheels made contact. "Touchdown!" said the pilot.

Nine seconds later, the nose wheels settled down, too. As *Columbia* rolled to a stop, a band struck up "The Star Spangled Banner" and from Mission Control at Johnson Space Center in Houston came the message, "Welcome home, *Columbia*. Beautiful. Beautiful."

*Columbia* signaled a bright future for the space program. It offered many commercial possibilities. It could, for example, be used to transport communications satellites into space, and at a cost that would be far less than launching them by rocket. The shuttle could even carry several satellites aloft at one time.

It could make possible many different types of scientific research. "The two great cosmic mysteries are the origin of the universe and the origin of life," said physicist Robert Jastrow. "The shuttle will give us a chance to probe both." No one doubted a new era in space investigation and exploration had begun.

# Astronaut Training

When it comes to working inside the orbiter, crew members have different duties and responsibilities. The commander or pilot operates and manages the orbiter. Payload specialists have assignments concerning the cargo the orbiter is carrying. They may, for example, deploy satellites from the cargo bay into space. To move objects into and out of the cargo bay, they use a 50-foot-long mechanical arm known as a remote manipulator system (RMS). Also aboard the orbiter may be a mission specialist who conducts scientific experiments.

Every member of the flight crew gets special training. There are courses in science, mathematics, meteorology, navigation, physics, and computer operation.

There is also aircraft training, with an emphasis on landings, on gliding in at high speed with no power — as the orbiter does.

Each candidate experiences zero gravity — weightlessness — aboard a specially equipped KC-135 aircraft. They float weightlessly inside the aircraft.

Training sessions are also held in a huge tank of water, called a "water im-

*Wearing a pressurized spacesuit, astronaut candidate Anna Fisher prepares to enter a water-filled tank to simulate the weightlessness of space.*

*Astronaut candidate Frederick Hauck floats freely during a period of weightlessness aboard a KC-135 aircraft.*

Astronaut candidate Guion (Guy) Bluford takes part in a training exercise to prepare him for a helicopter rescue following a parachute landing in water.

*Astronaut candidate Ronald McNair (left) gets instruction concerning the controls of a T-38 aircraft from fellow trainee S. David Griggs.*

mersion facility." Being immersed in water is something like being weightless in space.

Classroom study is also important. The trainees learn about the design and function- of launch vehicles, about shuttle housekeeping and emergency procedures, and about shuttle guidance and navigation.

The training gets exciting when the candidates start working with the various simulators. Each simulator is a full-scale model of the orbiter and its controls.

The simulator's instruments can be programmed to give typical readings concerning the orbiter's speed and direction.

At the same time, a screen shows views of the launchpad or the landing runway, the earth or the stars. After a training session in a simulator, the candidate feels almost as if he or she has completed a real flight.

Even after the trainees have been accepted into the program and assigned to a flight crew, the schooling continues. They then train with other members of their crew. And in the days and weeks before their actual flight, the crew and flight controllers practice an entire mission — from launch to landing — in a joint training exercise.

# First U.S. Spacewoman

In mid-June, 1983, beachfront motels near the Kennedy Space Center in Florida put out SORRY, NO VACANCY signs. More than 1,500 journalists packed the press grandstand. Television crews readied their cameras and sound equipment. On the white-sand beaches, more than a half-million people watched. Not since the first flight of *Columbia* some two years before had a space shuttle lift-off caused so much excitement.

The reason for all the interest was not the shuttle *Challenger* itself; it was one of the crew members — Sally Kristen Ride.

When *Challenger* headed off on the seventh shuttle mission, Sally Ride became the first American woman to travel in space. The mission, which also included four male astronauts, lasted six days.

Ride, described as a "normal, healthy person who loves sports," was born in Los Angeles. In 1978, while studying for a Ph.D. degree in physics from Stanford University, she applied to join the space program.

Shortly after, she became an astronaut candidate. In 1982, a year before her first flight, she married astronaut Steven Hawley.

*As an astronaut candidate, Ride gets positioned in an aircraft-type seat for training in ejection from an aircraft.*

*Sally Ride, the first American woman in space.*

*Ride and other crew members leave astronaut quarters to go to the launch complex for the historic flight. Besides Ride, the crew includes Robert Crippen, Frederick Hauck, Dr. Norman Thagard, and John Fabian.*

During the flight, Ride's chief task was scheduled on the fifth day in space. She and astronaut John Fabian, an Air Force colonel, used the 50-foot-long "cherry picker," the remote-controlled arm, to pluck a satellite out of the *Challenger*'s cargo bay and release it into space.

While Ride called out commands, Fabian punched buttons on a console that guided the arm's movements.

The excitement over Ride's flight all but obscured another important first established by the mission. It was the first to carry five crew members, one more than any other previous flight.

It was also the first shuttle scheduled to glide to a landing on the new three-mile-long concrete runway at Kennedy Space Center. But low-hanging clouds over Cape Canaveral forced the *Challenger* to land in California's Mojave Desert, where five previous shuttles had come home.

Ride made a second shuttle flight in 1984. On that mission, she used the remote manipulator to launch a satellite that was used to measure the sun's effect on the earth's weather.

But it is for her first flight that Sally Ride will be remembered. As the first

American woman in space, she gave a big boost to the shuttle program.

Ride seemed to enjoy herself thoroughly during the mission. She took part in a contest to see which of the astronauts could float the fastest through the cabin. (She finished second.) She helped retrieve jelly beans that were floating around the ship. Afterward, she said, "I'm sure it's the most fun I'll ever have in my life."

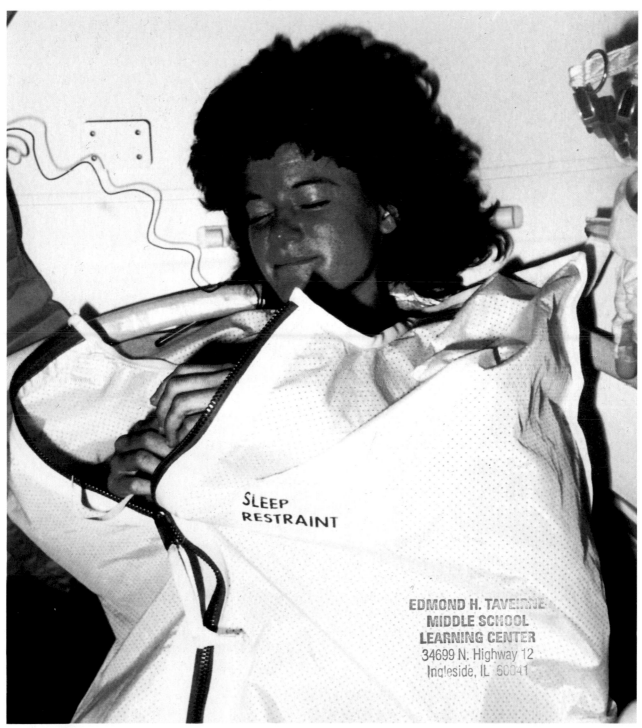

SLEEP
RESTRAINT

EDMOND H. TAVEIRNE
MIDDLE SCHOOL
LEARNING CENTER
34699 N. Highway 12
Ingleside, IL 60041

*Sally Ride photographed at her sleep station during* Challenger's *flight.*

# Shuttle Missions

In February, 1984, shortly after *Palapa* and *Westar* satellites were launched from the shuttle *Challenger*, something went wrong with the rockets that were supposed to boost them into orbit. Stranded in space, the two drum-shaped satellites, worth about $35 million apiece, were useless.

Almost immediately, space officials began planning a rescue mission. Astronauts would be sent into space aboard the shuttle *Discovery* to round up the two satellites and bring them back to Earth to be repaired.

By November, 1984, all was in readiness. Bad weather caused a one-day delay of the shuttle's launch. But once airborne, *Discovery* performed without any problems. The four-man, one-woman crew prepared for the salvage mission.

On their sixty-sixth orbit of the earth, the astronauts spotted *Palapa*. "The sun is up," said orbiter pilot David Walker. "We're ready to go!"

In bulky space suits, astronauts Joseph Allen and Dale Gardner glided out to meet the satellite. Gardner hooked himself into a pair of footholds on the shuttle's hull. Allen, using thrusters in his backpack, moved close to the rotating cylinder.

Allen carried a "stinger" spear, a long

*Astronaut Joseph Allen, wielding a "stinger" spear, approaches the* Westar *satellite.*

pole mounted on a round base. Allen's objective was to fire the stinger into the satellite's rear motor. An expanding probe would then lock into the satellite. "It's like opening an umbrella inside a chimney," Allen explained.

Once he had speared the satellite, Allen declared, "I've got it tied," and he began rocketing back to the mother ship with it.

As Allen held the satellite in place, Gardner removed the stinger. Then the two men wrestled the satellite into *Discovery*'s cargo bay. The next day, the crew retrieved *Westar*, the other satellite.

The mission was hailed as one of the most successful in the history of the space program. President Ronald Reagan led the cheers and applause. "You have demonstrated that we can work in space in ways we never imagined were possible," he said.

With the two satellites locked in the

*Allen gets in close to the satellite and prepares to capture it for return to Earth in* Discovery's *cargo bay.*

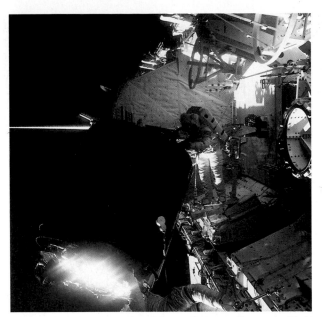

*Astronauts Joseph Allen (bottom) and Dale Gardner struggle with the captured* Palapa *satellite in an effort to secure it in* Discovery's *cargo bay.*

cargo bay, *Discovery* headed back to a smooth landing at the Kennedy Space Center.

Salvaging satellites that have gone astray is only one of the uses to which the shuttle has been put. There are many others. For example, in November, 1983, in what was the ninth shuttle flight, *Challenger* carried a scientific research station into orbit. Called *Spacelab*, it represented a big advance over *Skylab*, the work station that had been placed in orbit ten years before (pages 30–33).

*Spacelab* was built in West Germany by the European Space Agency, a scientific organization made up of Western European nations. It consists of two principal parts: a manned space laboratory and several separate platforms called "pallets."

In the pressurized laboratory, scientists work in shirtsleeves in an Earth environment, conducting experiments in the manufacture of medicines and in the production of biological materials.

The V-shaped pallets outside the lab serve as platforms for carrying instruments used for experiments in astronomy and other fields. The equipment includes radar units, antennas, and telescopes.

In its first ten days of use, some 70 experiments were conducted aboard *Spacelab*. The orbital laboratory has been designed to have a lifetime of several years. Scientists have predicted that *Spacelab* research and experiments will make for significant advances in atmospheric studies, astronomy, solar physics, and biology.

The shuttle is capable of trucking even very heavy payloads into space. The shuttle *Challenger*, for instance, once lifted a huge Tracking and Data Relay Satellite into orbit. At 5,000 pounds, it was, at the time, the heaviest object ever carried aloft by a shuttle.

Space officials plan hundreds more shuttle missions. In humanity's effort to explore and conquer space, no one doubts the shuttle will continue to play the leading role for the remainder of the century and even beyond.

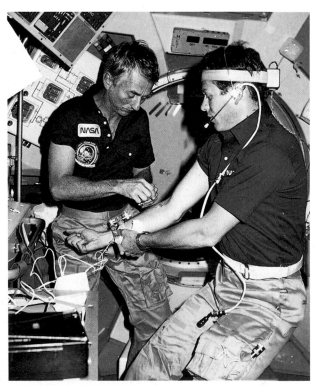

*Aboard* Spacelab, *mission specialist Owen Garriott draws blood from payload specialist Byron Lichtenberg for later testing on Earth.*

*Artist's drawing of shuttle's robotlike device (called a remote manipulator system) that is used for lifting satellites in and out of cargo bay.*

# National Tragedy

By the end of 1985, American astronauts had soared into space 55 times. The shuttle had made 24 successful flights, and had come to be looked upon as a workhorse. People had begun to take space travel for granted.

Then tragedy struck, and Americans were shocked out of their ho-hum attitude toward space flight.

On January 28, 1986, the shuttle *Challenger* exploded shortly after lift-off. Its seven crew members disappeared in an orange-and-white fireball 9 miles above the Atlantic Ocean off Cape Canaveral.

Millions of Americans watched on television as the terrible sight was replayed again and again. They were shocked into silence and tears.

The tragedy was deepened because the crew had included Christa McAuliffe, a wife, a mother of two small children, and a popular high school teacher from Concord, New Hampshire. She was to be the first ordinary citizen shot into space. She had been in training for three months.

While aboard the *Challenger*, one of McAuliffe's tasks would have been to conduct two 15-minute classes as millions of schoolchildren watched by means of closed-circuit TV. One of the classes was to be based on a tour of the orbiter. "The ultimate field trip," she called it.

In the days that followed the terrible accident, Americans were united in mourning. Flags were lowered to half-staff. President Reagan, addressing the nation in a special broadcast, said: "We will never forget them nor the last time we saw them . . . as they prepared for their journey and waved good-bye and 'slipped the surly bonds of Earth to touch the face of God.' "

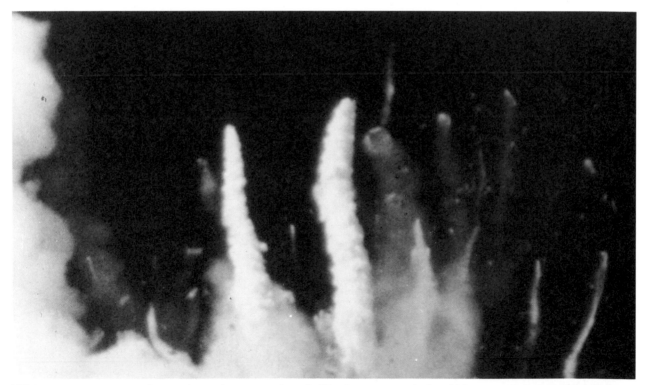

*With seven crew members aboard, space shuttle* Challenger *explodes 74 seconds after lift-off from Kennedy Space Center.*

"All the News That's Fit to Print"

# The New York Times

**Late Edition**

Weather: Partly cloudy and cold today, chance of snow; chance of snow tonight. Partly cloudy, cold tomorrow. Temperatures: today 27-30, tonight 13-19; yesterday 14-23. Details, page C19.

VOL.CXXXV . . . No. 46,669  NEW YORK, WEDNESDAY, JANUARY 29, 1986 — except on Long Island — 30 CENTS

# THE SHUTTLE EXPLODES

## 6 IN CREW AND HIGH-SCHOOL TEACHER ARE KILLED 74 SECONDS AFTER LIFTOFF

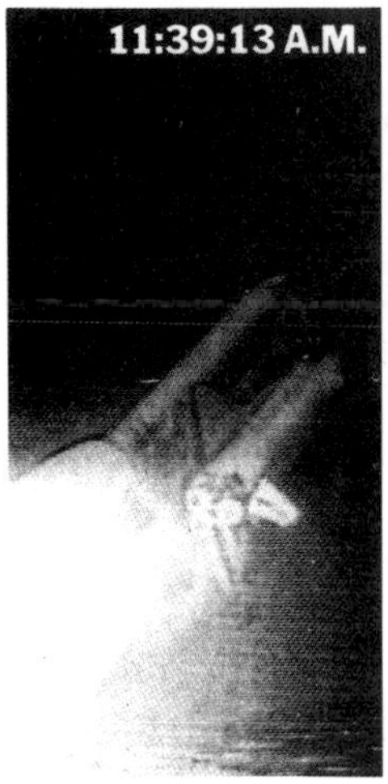
11:39:13 A.M.

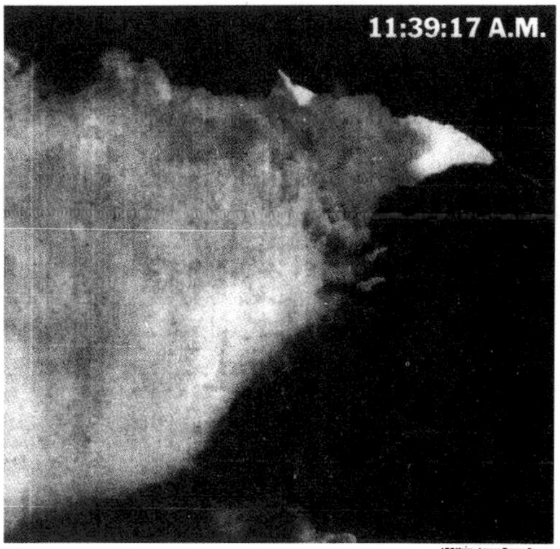
11:39:17 A.M.

ABC News; Agence France-Presse

## Thousands Watch A Rain of Debris

By WILLIAM J. BROAD

Special to The New York Times

CAPE CANAVERAL, Fla., Jan. 28 — The space shuttle Challenger exploded in a ball of fire shortly after it left the launching pad today, and all seven astronauts on board were lost.

The worst accident in the history of the American space program, it was witnessed by thousands of spectators who watched in wonder, then horror, as the ship blew apart high in the air.

Flaming debris rained down on the Atlantic Ocean for an hour after the explosion, which occurred just after 11:39 A.M. It kept rescue teams from reaching the area where the craft would have fallen into the sea, about 18 miles offshore.

It seemed impossible that anyone could have lived through the terrific explosion 10 miles in the sky, and officials said this afternoon that there was no evidence to indicate that the five men and two women aboard had survived.

### No Ideas Yet as to Cause

There were no clues to the cause of the accident. The space agency offered no immediate explanations, and said it was suspending all shuttle flights indefinitely while it conducted an inquiry. Officials discounted speculation that cold weather at Cape Canaveral or an accident several days ago that slightly damaged insulation on the external fuel tank might have been a factor.

Americans who had grown used to the idea of men and women soaring into space reacted with shock to the disaster, the first time United States astronauts had died in flight. President Reagan canceled the State of the Union Message that had been scheduled for tonight, expressing sympathy for the families of the crew but vowing that the nation's exploration of space would continue.

Killed in the explosion were the mission commander, Francis R. (Dick) Scobee; the pilot, Comdr. Michael J. Smith of the Navy; Dr. Judith A. Resnik; Dr. Ronald E. McNair; Lieut. Col. Ellison S. Onizuka of the Air Force; Gregory B. Jarvis, and Christa McAuliffe.

Mrs. McAuliffe, a high-school teacher from Concord, N.H., was to have been the first ordinary citizen in space.

### After a Minute, Fire and Smoke

The Challenger lifted off flawlessly this morning, after three days of delays, for what was to have been the 25th mission of the reusable shuttle fleet that was intended to make space travel commonplace. The ship rose for about a minute on a column of smoke and fire from its five engines.

Suddenly, without warning, it erupted in a ball of flame.

The shuttle was about 10 miles above the earth, in the critical seconds when the two solid-fuel rocket boosters are firing as well as the shuttle's main engines. There was some discrepancy about the exact time of the blast: The National Aeronautics and Space Administration said they lost radio contact with the craft 74 seconds into the flight, plus or minus five seconds.

Two large white streamers raced away from the blast, followed by a rain of debris that etched white contrails in the cloudless sky and then slowly

Continued on Page A5, Column 4

## From the Beginning to the End

*The last flight of the shuttle Challenger lasted about 74 seconds. Here is the transcript, as recorded by The New York Times, of its final moments, before and after liftoff.*

PUBLIC AFFAIRS OFFICER: Coming up on the 90-second point in our countdown. Ninety seconds and counting. The 51-L Mission ready to go. . . .

T minus 10, 9, 8, 7, 6, we have main engine start, 4, 3, 2, 1. And liftoff. Liftoff of the 25th space shuttle mission and it has cleared the tower. . . .

MISSION CONTROL CENTER: Watch your roll, Challenger.

PUBLIC AFFAIRS OFFICER: Roll program confirmed. Challenger now heading down range. [Pause.] Engines beginning throttling down now at 94 percent. Normal throttle for most of flight 104 percent. Will throttle down to 65 percent shortly. Engines at 65 percent. Three engines running normally. Three good cells, three good ABU's. [Pause.] Velocity 2,257 feet per second, altitude 4.3 nautical miles, down range distance 3 nautical miles. [Pause.]

Engines throttling up, three engines now at 104 percent.

MISSION CONTROL: Challenger, go with throttle up.

FRANCIS R. SCOBEE, CHALLENGER COMMANDER: Roger, go with throttle up.

PUBLIC AFFAIRS OFFICER: One minute 15 seconds, velocity 2,900 feet per second, altitude 9 nautical miles, down range distance 7 nautical miles. [Long pause.]

Flight controllers here looking very carefully at the situation. [Pause.]

Obviously a major malfunction. We have no downlink [communications from Challenger]. [Long pause.]

We have a report from the flight dynamics officer that the vehicle has exploded.

## *How Could It Happen? Fuel Tank Leak Feared*

By MALCOLM W. BROWNE

Debris from the explosion of the shuttle Challenger was scattered so widely over the Atlantic Ocean that investigators may never recover enough of it to pin down the cause of the disaster. But suspicions quickly focused on the craft's huge external fuel tank, a potential bomb that carried more than 385,000 gallons of liquid hydrogen and more than 140,000 gallons of liquid oxygen at liftoff.

The most logical explanation is that a large leak must have occurred either in the tank itself or in the pipeline and pumping system that carried liquid hydrogen to the orbiter's three main engines.

Barbara Schwartz, a spokesman for the Johnson Space Center, acknowledged that pure liquid or gaseous hydrogen cannot burn; only if the pure hydrogen carried in the rear section of the shuttle's tank were allowed to come into contact with air, or with the liquid oxygen in the tank's nose section, could it have burned or exploded.

### Potential Dangers of Hydrogen Gas

But what might have started the leak, and what could have ignited the explosion that followed?

Parallel questions, never fully answered, were raised after the fire that destroyed the German airship Hindenburg as it was landing at Lakehurst, N.J., on May 6, 1937. The shuttle Challenger, like the Hindenburg, had been releasing hydrogen gas into the air shortly before the disaster, and some of the gas might have remained aboard the craft, mixed with air and ready to detonate if exposed to the smallest spark.

Neither NASA nor Martin Marietta Aerospace, the manufacturer of the external fuel tank, would comment yesterday on possible causes of the disaster.

But the geometry of the shuttle's external fuel tank, as described by official manuals from NASA and the Rockwell International Corporation, a major shuttle contractor, suggest one potential danger point in particular: the "intertank," or midsection of the structure, which separates the liquid oxygen tank from the liquid hydrogen tank. The bulk of the hydrogen fuel is closest to the liquid oxygen at this point, and a rupture or leak in the plumbing or walls of the intertank could have flooded the two fluids together to create a gigantic bomb.

Suggestions that the unseasonably cold weather at

Continued on Page A4, Column 1

## *After the Shock, a Need to Share Grief and Loss*

By SARA RIMER

The nation came together yesterday in a moment of disaster and loss. Wherever Americans were when they heard the news — at work, at school or at home — they shared their grief over the death of the seven astronauts, among them one who had captured their imaginations, Christa McAuliffe, the teacher from Concord, N.H., who was to have been the first ordinary citizen to go into space.

Shortly before noon, when the first word of the explosion came, daily events seemed to stop as people awaited the details and asked the same questions: "What happened? Are there any survivors?"

In offices, restaurants and stores, people gathered in front of television sets, mesmerized by the terrible scene of the shuttle exploding, a scene that would be replayed throughout the day and night. Children who had learned about Mrs. McAuliffe were watching in classrooms across the country.

It seemed to be one of those moments, enlarged and frozen, that people would remember and recount for the rest of their lives — what they were doing and where they were when they heard that the space shuttle Challenger had exploded. The need to reach out, to speak of disbelief and pain, was everywhere. Family members telephoned one another, friends telephoned friends.

"It was like the Kennedy thing," said John Hannan, who heard the news when his sister called him at his office, a personnel recruiting concern in Philadelphia. "Everyone was numb."

### 'I Felt Very Close to Her'

Florine Israel, a legal secretary at the New York Civil Liberties Union, echoed the sentiments of many who spoke of Mrs. McAuliffe not as an astronaut but as a friend. "I felt very close to her," she said. "She was ordinary people. She was a mother, a working woman. I felt like I was a part of it."

The image of the shuttle exploding flashed across 100 television sets in the electronics department of Macy's, in midtown Manhattan, where a crowd of workers from nearby offices and facto-

Continued on Page A3, Column 1


| | | | |
|---|---|---|---|
| About New York | B24 | Movies | C18 |
| Around Nation | A11 | Music | C15,C20,C24 |
| Books | C17 | Obituaries | A21 |
| Bridge | C20 | Op-Ed | A23 |
| Business Day | D1-22 | Real Estate | D22 |
| Crossword | C20 | Sports Pages | B5-9 |
| Day by Day | B3 | Theaters | C13,C18 |
| Editorials | A22 | TV / Radio | C20-21,C23 |
| Going Out Guide | C18 | U.N. Events | A14 |
| Letters | A22 | Washington Talk | A20 |
| Living Section | C1-12 | Weather | C19 |

**News Summary and Index, Page B1**

*Classified Ads* B12-23 | *Auto Exchange* B9-12


## *Reagan Lauds 'Heroes'*

President Reagan, shaken by the explosion of the space shuttle, postponed his State of the Union Message. "We mourn seven heroes," he said in a talk broadcast from the White House after the disaster. "There will be more shuttle flights and more shuttle crews and, yes, more volunteers, more civilians, more teachers in space."

He also sought to console the nation's pupils, many of whom saw telecasts of the loss of the teacher who was to have been sent into space. Article and transcript, page A9.

A B C D E F

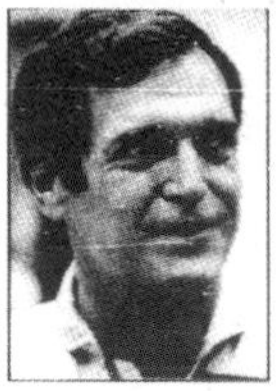

**Francis R. Scobee**
*Commander*

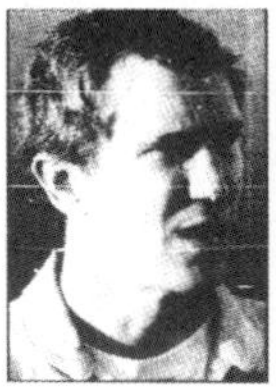

**Michael J. Smith**
*Pilot*

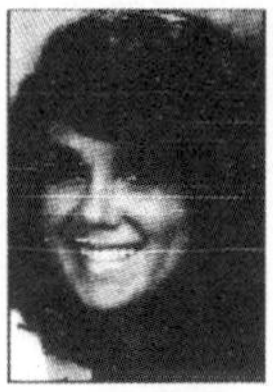

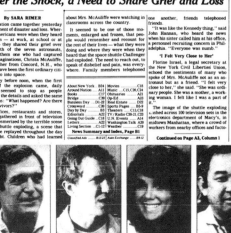

**Judith A. Resnik**
*Electrical Engineer*

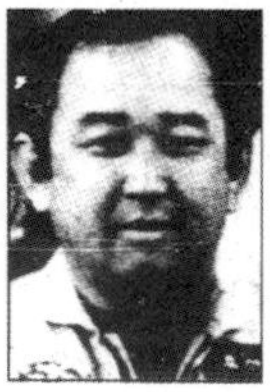

**Ellison S. Onizuka**
*Engineer*

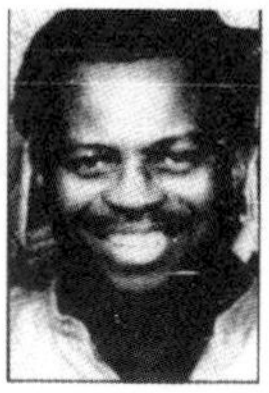

**Ronald E. McNair**
*Physicist*

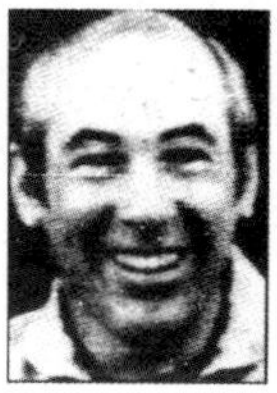

**Gregory B. Jarvis**
*Electrical Engineer*

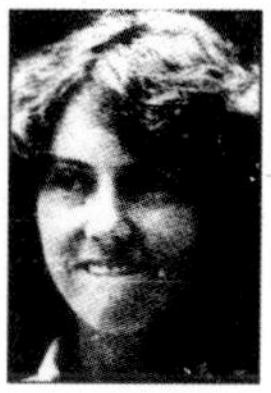

**Christa McAuliffe**
*Teacher*

Crew members of shuttle Challenger *that exploded on January 28, 1986 (from left): payload specialists Christa McAuliffe and Gregory Jarvis, mission specialist Judith Resnik, commander Francis Scobee, mission specialist Ronald McNair, pilot Michael Smith, and mission specialist Ellison Onizuka.*

# Back in Space

After the *Challenger* disaster, all shuttle flights were halted. No one doubted that the three remaining orbiters — *Columbia*, *Discovery*, and *Atlantis* — would fly again. But their launching would have to wait until the flaw that caused the *Challenger* tragedy could be identified and corrected.

President Reagan appointed a commission to investigate the accident. Late in 1986, the commission reported that the explosion was caused by the failure of synthetic rubber rings that were supposed to seal a joint in the shuttle's right rocket booster. Called O-rings, they had permitted hot gases to leak out of the booster. The gases then burned a hole in the shuttle's external fuel tank, which triggered the explosion. Unusually cold weather on the morning of the flight, plus

*Leaving the orbiter* Discovery, *Commander Rick Hauck and his crew are welcomed back by the then Vice President George Bush. Other crew members include (from top): mission specialists George Nelson, David Hilmers, and John Lounge, and pilot Richard Covey.*

"All the News That's Fit to Print"

# The New York Times

**Late Edition**
New York: Today, mostly sunny and milder. High 68-72. Tonight, clear, not as cool. Low 55-61. Tomorrow, partly cloudy and warmer. High 75-79. Yesterday: High 67, low 49. Details, page C30.

VOL.CXXXVIII . . . No. 47,844 — NEW YORK, FRIDAY, SEPTEMBER 30, 1988 — 35 CENTS

# AMERICAN ASTRONAUTS ROAR BACK TO SPACE, RENEWING THE NATION'S HOPES FOR SHUTTLE

## *U.N. Peacekeeping Forces Named Winner of the Nobel Peace Prize*

By SHEILA RULE
Special to The New York Times

OSLO, Sept. 29 — The United Nations peacekeeping forces, which for 40 years have been deployed to reduce tensions in the world's trouble spots, were named here today as the 1988 winner of the Nobel Peace Prize.

The Norwegian Nobel Committee said the peacekeeping forces were being recognized because they "represent the manifest will of the community of nations to achieve peace through negotiations and the forces have, by their presence, made a decisive contribution toward the initiation of actual peace negotiations."

**Recognizing Expanded Role**

The award comes at a time when the United Nations, long criticized as a bloated, ineffectual and irrelevant organization, has been experiencing a renaissance. Students of the organization say that, perhaps more than at any time in its 40-year history, the United Nations is involved in the sort of peace initiatives for which it was founded.

Egil Aarvik, chairman of the Norwegian Nobel Committee, said he hoped that the United Nations Secretary General, Javier Pérez de Cuéllar, would come to Oslo to accept the prize at a ceremony here Dec. 10, the anniversary of Alfred Nobel's death in 1896. The prize consists of a diploma and gold medal and a cash prize of 2.5 million Swedish kronor, or about $388,000.

"It is the considered opinion of the committee," the selection committee said in a written statement released when the award was announced, "that the peacekeeping forces through their efforts have made important contributions toward the realization of one of the fundamental tenets of the United Nations. Thus the world organization has come to play a more central part in world affairs and has been invested with increasing trust."

Thousands of peacekeeping forces, which are recruited from many nations, are currently stationed along the India-Pakistan border, in the Sinai,

*Continued on Page A10, Column 1*

## ARBITRATORS RULE IN FAVOR OF EGYPT ON SINAI ENCLAVE

### Israeli Withdrawal Called for but Small Area of Border Is Left for Negotiation

By YOUSSEF M. IBRAHIM
Special to The New York Times

GENEVA, Sept. 29 — An international arbitration panel ruled in Egypt's favor today in a dispute with Israel over a border enclave, but left it to the two sides to work out a final section of the boundary.

The border dispute has been the major unresolved issue between the two nations since they completed a peace accord in 1979.

Egypt had insisted that Israel withdraw from the stretch of land, Taba, which covers about 250 acres along the Gulf of Aqaba. Although the disputed territory was small, the issue stirred strong emotions in both Israel and Egypt, with each insisting the other was unreasonable in seeking the land.

**Occupied in 1967 War**

The area was occupied by Israel in the 1967 war when it captured the Sinai Peninsula. If Israel withdraws from Taba, Egypt would return to the borders that existed before the war.

The dispute was complicated when Israelis built a beach resort in the area and put pressure on the Israeli Government not to yield the enclave.

Israel had formally maintained that the border drawn between Egypt and Palestine in 1906 put Taba inside what is now its territory, thus allowing the resort to remain in Israeli hands. The two nations failed to resolve the dispute amicably and agreed to binding arbitration two years ago.

While the Israeli judge on the five-

*Continued on Page A11, Column 1*

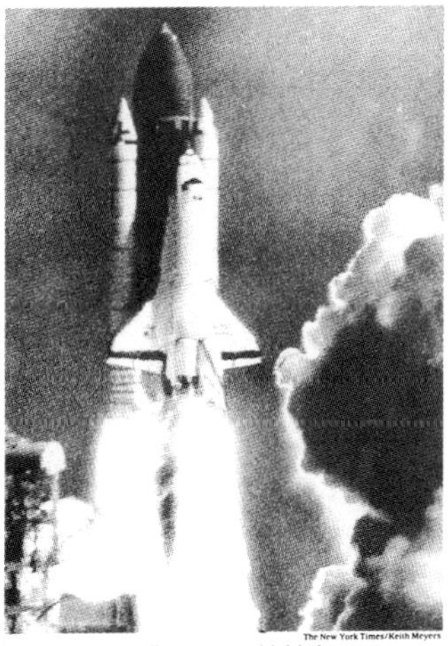

The New York Times/Keith Meyers
The Discovery lifting off to end a 32-month halt in the space program.

## LONG HIATUS ENDS

### NASA Chief Sees Liftoff as First Step of New Era After Disaster

By JOHN NOBLE WILFORD
Special to The New York Times

CAPE CANAVERAL, Fla., Friday, Sept. 30 — The space shuttle Discovery, picking up the baton from the fallen Challenger, raced into orbit Thursday carrying five experienced astronauts and the nation's hopes for revival of its civilian space program.

The successful launching ended 32 months of gloom, self-doubt and redesign following the worst disaster in the history of space flight.

With dark memories of the Challenger explosion shadowing their every step, space agency officials pressed ahead with the tense and frequently interrupted countdown. They wrung their hands over wind conditions aloft, but after hours of analysis and deliberation, they finally gave the "go" command for the planned four-day mission, waiving a rule against launching in unfavorable wind conditions in their determination to return to space.

**Stunning Moment Recalled**

Liftoff came at 11:37 A.M. Thursday, one hour and 38 minutes behind schedule. The Discovery soared into the sky on the orange flame and billowing white vapors of its rockets. It disappeared into the high clouds, but the rolling thunder of exhaust continued past the 73-second mark that was on everyone's mind. That was the stunning moment on Jan. 28, 1986, when a booster on the Challenger failed and the craft exploded, killing all seven astronauts and grounding the nation's principal vehicles for going into space.

When the two redesigned solid-fuel boosters completed firing, and again when the Discovery finally reached orbit, the tension in the firing control room turned to relief and then excitement. Exuberant controllers abandoned professional reserve and let out whoops and cheers. The "emotional high" in the control room was something officials said they had never seen on any previous launching.

Addressing the jubilant launching team, James C. Fletcher, the Administrator of the National Aeronautics and Space Administration, said: "It's been a long wait. The nation owes you a lot. This is the first step of a new era."

**Boosters Are Flawless**

The solid-fuel boosters, blamed for the Challenger disaster, are like the craft itself among the flight's experiments and were continuously monitored. They performed without apparent flaw. [Page D22.]

All other major systems were reported to be functioning satisfactorily. And five American astronauts were safely orbiting 184 miles above the earth, completing a revolution once every 90 minutes. The men on Discov-

*Continued on Page D22, Column 3*

## *Shouts, Tears and Applause Amid a Vast Wave of Relief*

By WILLIAM E. SCHMIDT

Gathered anxiously in front of their television screens, millions of Americans watched the space shuttle Discovery thunder spectacularly into orbit yesterday morning amid applause and cheers but, most of all, amid an overwhelming sense of relief.

The launching of the Discovery offered Americans more than renewed pride; it provided a kind of national exorcism, an opportunity to purge the doubts and the uncertainty about the nation's prowess and technological ability that had lingered since the explosion of the shuttle Challenger in 1986.

**'I Had My Fingers Crossed'**

In Washington, a beaming President Reagan told an audience in the Rose Garden, "America is back in space." But he also confessed that during the launching, "I think I had my fingers crossed like everybody else."

On the campaign trail, Michael S. Dukakis and George Bush found common ground, for once, in the Discovery's success. "We're going to keep the edge in space," Mr. Bush said in St. Charles, Mo. "America's back." Mr. Dukakis, in New Jersey, called it a "very successful morning," adding, "We're very proud of the astronauts."

In Nashville, Senator Lloyd Bentsen of Texas, the Democratic Vice-Presidential nominee, delayed a campaign flight to watch the launching on a borrowed black and white television. "O.K., Russians, here we come," he declared as the Discovery rose off the pad.

**Tears and Melancholy**

For others, the exhilaration was tinged with tears, sadness and the melancholy of remembrance.

At Engine Company 98, north of downtown Chicago, Capt. Edmund J. Enright of the Fire Department stood in the firehouse kitchen, watching the television screen as the Discovery rode a pillar of flame. "You know," he said quietly, "those other ones they lost, I bet they're in there riding with them."

In Brigham City, Utah, where workers for Morton Thiokol designed and built the shuttle's solid-fuel booster rockets, Mayor Peter Knudson cheerfully described the launching as "one of the greatest moments of my life." It was a faulty joint on one of Challenger's 126-foot booster rockets that was blamed for the explosion.

In Pasadena, Calif., at the Jet Pro-

*Continued on Page D23, Column 1*

## Fiscal Experts Give Dukakis More Credit Than Bush Does

By ALLAN R. GOLD
Special to The New York Times

BOSTON, Sept. 29 — Vice President Bush takes glee in terming "the Massachusetts miracle," a slogan adopted by Gov. Michael S. Dukakis for the state's economic upturn, "the Massachusetts mirage." He calls his Democratic opponent the nation's leading taxer, spender and borrower, saying that Mr. Dukakis is presiding over the erosion of his state's manufacturing base.

But finance experts agree that under Governor Dukakis the Massachusettts economy and budget have been considerably stronger than Mr. Bush suggests.

Even before Mr. Bush began trying to raise doubts among voters about Mr. Dukakis's governing and his ability to manage fiscal and economic affairs, which Mr. Dukakis considers a strong suit, Mr. Dukakis was fending off questions at home about the state's budget problems. Also, the extent to which Mr. Dukakis can take credit for a vibrant economy has often been disputed.

**Ignores Broader Trends**

Taken at face value, many of Mr. Bush's assertions about how Governor Dukakis has managed fiscal and economic affairs are accurate. But his charges often rely on selected information interpreted in the narrowest, most negative way possible, ignoring contradictory evidence and broader trends that show that in terms of personal income and unemployment, Massachusetts has had one of the strongest economies of any state.

"I do get ticked off with his off-the-wall stuff that is just not true about the state," Richard A. Manley, president of the Massachusetts Taxpayers Foundation, said of Mr. Bush's remarks. The

*Continued on Page B7, Column 1*

## *Citing Drug Use, Olympic Official Proposes a Ban on Weight Lifting*

By MICHAEL JANOFSKY
Special to The New York Times

SEOUL, South Korea, Friday, Sept. 30 — An Olympic official said Thursday that he would recommend that weight lifting be dropped from future Olympic Games because of drug use by some athletes.

Richard Pound, vice president of the International Olympic Committee, disclosed his plan after a fifth weight lifter was disqualified from the Games for testing positive for a banned drug.

Among eight athletes disqualified at the Olympics after their drug tests produced positive results, five were weight lifters, three of them medal winners.

In addition, the British Olympic Committee acknowledged today that two British athletes, one in track and field, the other in judo, had tested positive. Their names were not released pending testing of their second samples and announcement of any disciplinary action.

Mr. Pound said that he believed the use of steroids was "endemic" among weight lifters and that the international federation governing the sport had not done enough to police their use. As a result, he said he would introduce his proposal at the I.O.C.'s next executive board meeting, in Vienna in December.

"Now may be the time to give weight lifting an Olympic holiday," Mr. Pound said. "There certainly seems to be a problem. Maybe we can take the sport to task, that until they clean themselves up, they can't get back in the

*Continued on Page A20, Column 6*

**Olympic Highlights**

**TRACK AND FIELD** Two sisters-in-law, Jackie Joyner-Kersee and Florence Griffith Joyner, each won a second gold medal. Kersee became the first American woman to win the Olympic long jump. Joyner set a world record in the 200-meter sprint.

**BASKETBALL** The Soviet Union captured the gold medal with a 76-63 victory over Yugoslavia. The U.S. won the bronze.

**TENNIS** Miloslav Mecir of Czechoslovakia won the gold medal in men's singles by beating Tim Mayotte of the United States in four sets.

*Articles begin on page A15.*

JEWISH WOMEN GIRLS REMEMBER TO LIGHT SHABBAT CANDLES 18 MINUTES BEFORE SUNSET. IN NYC 6:32 P.M. INFO (718) 774-2026 ADVT.

SEE TAMARA TONIGHT! AD IN A&E Section (212) 239-2500 ADVT.

MELISSA GRAHAM MURPHY—I WILL LOVE you forever. Marry me. "Your Boy." ADVT.

## Charles Addams Dead at 76; Found Humor in the Macabre

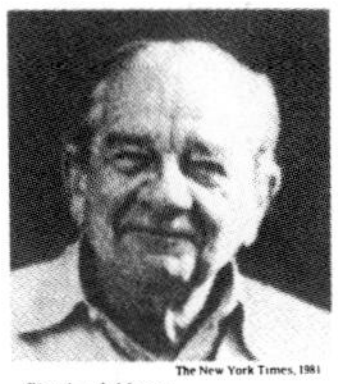

The New York Times, 1981
Charles Addams

By ERIC PACE

Charles Addams, the cartoonist whose macabre humor brought a touch of ghoulishness to The New Yorker's glossy pages for five decades, died yesterday at St. Clare's Hospital and Health Center, 415 West 51st Street. Virginia Stuart, director of community relations at the hospital, said Mr. Addams died in the emergency room after being brought to the hospital by ambulance. He was 76 years old and had an apartment in midtown Manhattan and a house in Sagaponack, L.I.

Mr. Addams's wife, Marilyn, said he had a heart attack Thursday morning in his automobile while it was parked in front of their apartment building.

"He's always been a car buff, so it was a nice way to go," Mrs. Addams said.

**12 Books Published**

A typical Addams cartoon was the one that showed a weird-looking man waiting outside a delivery room with a nurse telling him, "Congratulations, it's a baby!"

Many others depicted a Frankensteinian butler, a slinkily witchlike mother and other odd denizens of a haunted-looking Victorian house. In one 1946 drawing, they are up in its tower, about to greet Christmas carolers by dousing them with what looks like boiling oil.

The New Yorker published its first cartoon by Mr. Addams in 1935, long before sick jokes and black humor came into vogue, and it remained the main showcase for his work. But his drawings were also collected in a suc-

*Continued on Page B8, Column 1*

**INSIDE**

**Feminists and the Law**
Scholars are redefining legal theory to reflect women's real-life perspec-
A tives. The Law, page B9.

B **Drugs Among the Police**
New York City has established a squad to look for drug-related crime
C at the precinct level. Page B1.

D **News Summary, Page A2**



| | | | |
|---|---|---|---|
| Bridge | C34 | Politics | B6-7 |
| Business Day | D1-21 | Real Estate | A28 |
| Crossword | C32 | Sports | A14-24 |
| Editorials | A34 | TV/Radio | C34-35 |
| Law | B9 | U.N. Events | A10 |
| Letters | A34 | Washington Talk | B5 |
| Media | D21 | Weather | C30 |
| Obituaries | B7-9 | Weekender Guide | C1 |
| Op-Ed | A35 | Word and Image | C33 |
| Classified Index | B10 | Auto Exchange | D26 |



E F G H

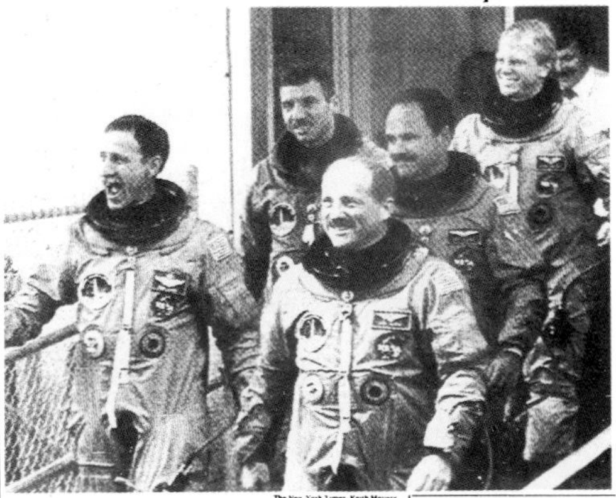

The New York Times/Keith Meyers
The Discovery crew heading for the spacecraft before the flight. In front were Col. Richard O. Covey of the Air Force, left, the pilot, and Capt. Frederick H. Hauck of the Navy, the mission commander. Behind them, from left, were Lieut. Col. David C. Hilmers of the Marine Corps, John M. Lounge and George D. Nelson.

THE NEW YORK TIMES is available for home or office delivery in most major U.S. cities. Please call this toll-free number 1-800-631-2500 ADVT

*Shuttle* Discovery *and its five-man crew are launched from Kennedy Space Center on September 29, 1988.*

design problems in the joint, had worsened the problem.

Almost three years passed before the shuttle was ready to fly again. Late in September, 1988, *Discovery* was successfully launched from its Kennedy Space Center pad, carrying five veteran astronauts aloft. As *Discovery* slowly rose, the space center's loudspeakers declared: "Americans return to space, as *Discovery* clears the tower."

While the spacecraft that went soaring into the sky looked much the same as those that had gone before, it was very different. The O-rings and booster joints had been redesigned, of course, but hundreds of other changes had also been made to increase the shuttle's safety and reliability.

In the orbiter itself, a new exit hatch had been installed to aid the crew in escaping in case of an emergency.

And for the first time since 1982, crew members wore space suits. Each was partly pressurized and equipped with an oxygen tank, a parachute, and an inflatable raft.

NASA scheduled 10 shuttle flights for 1990, 9 for 1991, and 13 for 1992. No one doubted that each of these would be conducted with far greater caution than existed that January morning in 1986 when *Challenger* suddenly exploded in the sunlit sky.

# The Future

Sometime in the near future, a shuttle will lift off from the Kennedy Space Center with an unusual payload in its cargo bay — the Hubble Space Telescope. A space observatory with a $1 billion price tag, the Hubble will enable scientists to look into the far corners of the universe.

The shuttle will release the Hubble, a cylinder that is about 44 feet long and 14 feet in diameter, into a permanent orbit some 300 miles above the earth. This puts it high above the earth's atmosphere.

The atmosphere frustrates earthbound astronomers by blocking out 90 percent of the energy from the stars. Hubble astronomers won't be confronted with this problem. They'll be able to look into the far reaches of space, perhaps unlocking age-old secrets about the universe: how it began, how it grew, how it is changing, and how those changes affect the earth. Many observers believe the Hubble (named for Edwin Hubble, an early American astronomer) to be the most important project that NASA has tackled in years.

What other space missions are planned for the years ahead?

A permanent manned space station is one major objective. *Skylab* was a space station, as is *Spacelab*. But neither of these was meant to be permanent.

What NASA plans to do is put a structure in space that will be kept there indefinitely, constructing it piece by piece from parts carried into orbit by shuttle flights.

Present plans call for the station to have dual keels, each 360 feet long. These are to be connected by upper and lower booms. Antennas and solar panels would be mounted on the booms. It is going to take several years to complete construction of the space station. Work

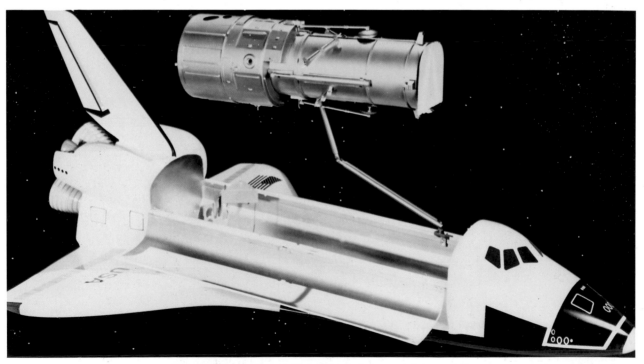

*The orbiter places a space telescope into orbit.*

*The space telescope will permit scientists to gaze farther into space than ever before.*

NASA's future plans include a permanent manned space station. This painting depicts an orbiter visiting such a station. The structure's cylinders contain the control center, living quarters, and laboratory. In other modules are life support systems, power units, and a communications system.

is scheduled to begin in the mid-1990s. The cost: about $28 billion.

The shuttle will be used to bring crews to the space station, and haul supplies such as food, oxygen, and the equipment and materials needed in experiments. Scientists will conduct research in the space station laboratory; there will be an observatory for astronomers. There might also be a manufacturing facility.

A permanent work station in space has been discussed since the early 1960s. The shuttle and its success has boosted interest in the idea. It may be a reality by the end of the century.

Another of NASA's goals is to put humans on the moon. This time, however, the space agency wants to keep them there and form a lunar colony. Thomas Paine, a former NASA official, has predicted that by the year 2025 the first humans will be calling themselves "natives of the moon."

In the distant future, the planet Mars could become the target of American astronauts. About half the diameter of the earth, Mars has a surface that is mostly a desert of reddish, rocky soil. The planet has an ice cap at both poles.

A good deal of what we know about Mars is based on information received from space probes. *Mariner 9*, launched in 1971, remained in Martian orbit for almost a year, sending back photographs and scientific information. Four immense volcanic mountains were found in the northern reaches of the planet, as well as a vast system of canyons and narrow channels. Some of the canyons are 50 to 75 miles wide and 3 to 4 miles deep.

Two other probes, *Viking 1* and *Viking 2*, landed on Mars in 1976. These landers had devices aboard that were able to perform chemical analyses of the Martian soil. They showed the presence of silicon, iron, calcium, and other minerals.

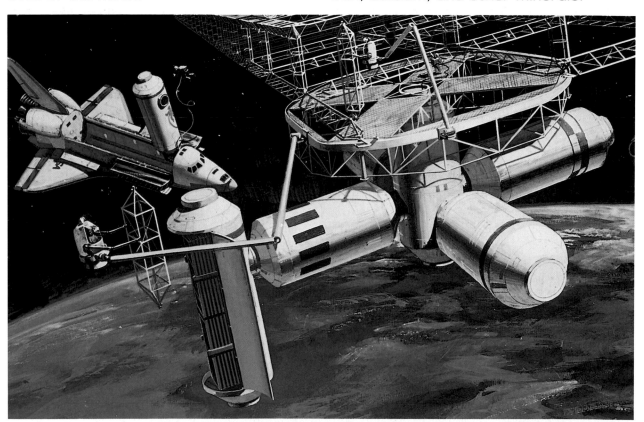

*Another concept of a manned space station. The cylinder-shaped modules, which would be delivered by orbiter, contain living quarters, laboratory, and support equipment.*

An outpost on the surface of the moon. In such a colony, water might be obtained by extracting it from the soil or by melting ice. A greenhouse would be built to grow green plants. Animals might be raised for protein. The soil could be processed to provide building materials — metals, glass, and cement.

A lunar mining operation. Ilmenite, which is fairly common in the lunar soil, is what is being mined. Ilmenite can be processed to produce water. Hydrogen and oxygen can then be derived from the water.

*An unmanned spacecraft nears touchdown on Mars. A parachute (at top) and rocket engines would slow the craft to a landing. After picking up soil samples, the ascent craft would fire its three-stage rocket motors to boost it out of the Martian atmosphere and on a return path to Earth.*

Information from other experiments showed, as one report put it, "the possibility of biological activity in the samples being incubated." Life on Mars is not likely, but it has not been positively ruled out.

Michael Collins, command pilot of the historic *Apollo 11* mission to the moon, and who has been with NASA since 1963, believes the United States space program should give priority to a Martian trip. "Inhospitable though it may be," says Collins in his book, *Liftoff*, "it's the closest thing to a second home we have in this solar system, with available oxygen and water. An expedition there is within the realm of possibility by the end of the century. . . .

"But," adds Collins, "we need to get started soon, or the world will pass us by."

Mars is the first planet beyond the earth, away from the sun. Even so, it is very far away. (Earth is located at an average distance of 93 million miles from the sun. Mars' average distance is 141 million miles from the sun, a difference of 48 million miles. The moon, by contrast, is only 238,850 miles from the earth.) A trip to Mars would take anywhere from twelve to thirty-six months. An astronaut crew bound for Mars would have to carry along many tons of food, water, and oxygen.

Despite the enormous problems to be overcome, a future generation may reach Mars and establish colonies there. As Robert Goddard, the famous American rocket scientist, who has been hailed as the "father of space flight," has put it: "The dream of yesterday is the hope of today and the reality of tomorrow."

# Important Dates in Space

## 1957

October 4 — *Sputnik 1*, the first artificial satellite is launched by the Soviet Union.

November 2 — Launching of *Sputnik 2*, which carried a dog named Laika, the first animal sent into space.

## 1958

January 31 — The United States launches its first satellite, *Explorer 1*.

July 28 — The National Aeronautics and Space Administration (NASA) is founded.

## 1959

December 4 — Recovery of chimpanzee Ham from *Mercury* capsule.

## 1961

April 12 — With the launching of *Vostok 1* by the Soviet Union, Yuri Gagarin becomes the first human in space; Gagarin orbits the earth.

May 5 — Cmdr. Alan B. Shepard, Jr., launched in a *Mercury 3* spacecraft, becomes the first American in space.

## 1962

February 20 — Lt. Col. John Glenn, in a *Mercury* spacecraft, becomes the first American to orbit the earth.

## 1963

June 16 — Valentina Tereshkova of the Soviet Union in *Vostok 6*, becomes the first woman to be launched into space.

## 1965

March 18 — Aleksei Leonov of the Soviet Union, in *Voskhod 2*, conducts the first space walk.

April 6 — Launching of *Intelsat 1*, the first communications satellite.

June 3 — Edward White, in *Gemini 4*, becomes the first American to walk in space.

## 1967

January 27 — Three American astronauts are killed in a cabin fire during a launchpad rehearsal at Kennedy Space Center.

## 1968

December 21–27 — *Apollo 8*, with Col. Frank Borman, Capt. James Lovell, Jr., and Lt. Col. William Anders, orbits the moon 10 times.

## 1969

July 16–24 — *Apollo 11* makes the first lunar landing. Neil Armstrong becomes the first human to walk on the moon.

## 1972

December 7–19 — *Apollo 17*, with Capt. Eugene Cernan, Cmdr. Ronald Evans, and Dr. Harrison Schmitt, makes last lunar mission. Cernan and Schmitt spend a record seventy-four hours and fifty-nine minutes on the moon's surface.

## 1973

May 14 — *Skylab* space station launched into Earth's orbit.

## 1975

July 15 — In joint United States–Soviet Union mission, *Apollo 18* and *Soyuz 19* link up in space.

August 20 — Launching of *Viking 1*, a probe to study the atmosphere and surface conditions of Mars.

September 9 — Launching of *Viking 2*, a second Mars probe.

## 1981

April 12–14 — First flight of space shuttle *Columbia*.

## 1982

February 8 — The Soviet Union launches *Salyut 7*, which sets space endurance record of 237 days.

## 1983

June 18 — Launching of space shuttle *Challenger*, carrying Sally Ride, the first American woman in space.

## 1984

August 30 — First flight of space shuttle *Discovery*.

## 1985

October 4 — First flight of space shuttle *Atlantis*.

## 1986

January 28 — Shuttle *Challenger* explodes 73 seconds after lift-off, killing all seven crew members.

## 1988

September 29 — United States space program resumes with launching of shuttle *Discovery*.

December 21 — Cosmonauts Vladimir Titov and Musa Manarov return to Earth after setting a new endurance record of 366 days spent in space.

## 1989

— Launching of space probe *Magellan* to Venus from space shuttle.

— Launching of space probe *Galileo* to Jupiter and Mars from space shuttle.

— Launching of Hubble Space Telescope from shuttle into Earth orbit.

# For Further Reading

## Young Readers

Branley, Franklyn. *From Sputnik to Space Shuttles*. New York, T. Y. Crowell, 1986.

Cross, Wilbur and Susanna. *Space Shuttle*. Chicago, Children's Press, 1985.

Fradin, Dennis. *Skylab*. Chicago, Children's Press, 1984.

Fradin, Dennis. *Space Colonies*. Chicago, Children's Press, 1985.

Friskey, Margaret. *The Moon-Ride Rock Hunt*. Chicago, Children's Press, 1972.

Taylor, L. B., Fr. *Space Shuttle*. New York, T. Y. Crowell, 1979.

## Young-Adult Readers

Allen, Joseph P. *Entering Space, An Astronaut's Odyssey*. New York, Stewart, Tabori & Chang, 1986.

Baker, Wendy. *America in Space*. New York, Crescent Books, 1986.

Collins, Michael. *Liftoff; The Story of America's Adventure in Space*. New York, Grove Press, 1988.

Mason, Robert Grant, Editor. *Life in Space*. Boston, Little Brown, 1983.

Osman, Tony. *Space History*. New York, St. Martin's Press, 1983.

Wilford, John Noble. *We Reach the Moon*. New York, Grosset & Dunlap, 1973.

# Index

Aldrin, Edwin, Jr., 8, 26—28
Allen, Joseph, 46—47
animals in space, 14, 64
Anders, William, 25, 65
Apollo Space Project, 8—10, 12, 25, 26—28
   *Apollo 8*, 25, 64
   *Apollo 11*, 8—10, 12, 26—28, 63, 64
   *Apollo 17*, 24, 28, 64
Armstrong, Neil, 8, 10, 26—28, 64
astronauts, U.S., 8—10, 17—18, 18—20, 24,
   25, 27, 28, 31, 42—45, 50, 60, 63, 64—
   65
   training, 38—41
*Atlantis* space shuttle, 35, 54, 65

booster joints, 54, 57
Borman, Frank, 25, 64

Cape Canaveral, 17, 18, 50
Cape Kennedy, 23
cargo bay, 38, 58
Cernan, Eugene, 28, 65
*Challenger* space shuttle, 35, 42—46, 48, 50,
   54, 65
   disaster, 50, 54, 57, 65
Collins, Michael, 8, 10, 26—28, 31, 63
*Columbia* space capsule, 27
*Columbia* space shuttle, 34, 35, 37, 42, 54, 65
commanders, 38
cosmonauts, 16—17, 22, 64

*Discovery* space shuttle, 35, 46, 54, 57, 65

*Eagle* (lunar module), 27—28
*Early Bird: see Intelsat 1*
earth orbit, 16—17, 18—20, 22, 24, 25, 26,
   30—33, 64, 65
endurance records, 65
European Space Agency, 48
Evans, Ronald, 28, 64
*Explorer 1*, 64
external fuel tanks, 35, 54

Fabian, John, 44
flight controllers, 41
flight simulators, 41
*Freedom 7* capsule, 17—18
*Friendship 7* capsule, 18, 20

Gagarin, Yuri, 16—17, 22, 64
*Galileo* Space Probe, 65
Gardner, Dale, 46—47

Gemini Space Project, 23—24, 64
Glenn, John, 18—20, 22, 64
Goddard, Robert, 63
gravity, 14, 26, 31, 38

Hubble Space Telescope, 58, 65

INTELSAT, 23
  *#1*, 23, 64

Jastrow, Robert, 37
Johnson Space Center, 27—28, 37
Jupiter, 65

Kennedy Space Center, 42, 44, 48, 57
Khruschev, Nikita, 17, 22

Laika, 14, 64
landing craft: *see* lunar modules
Leonov, Aleksei, 24, 64
Lovell, James, Jr., 25, 64
lunar colonies, 60
lunar landscape, 8—10
lunar modules, 25, 27—28
lunar surface, 8—10, 25

McDivitt, James, 23—24
*Magellan* Space Probe, 65
*Mariner* Space Probe, 60
Mars, 60—63, 65
Mercury Space Project, 14, 17—18, 18—20, 64
meteor showers, 33
Mission Control (Houston), 18, 24, 27—28, 37
mission specialists, 38
moon: *see also* lunar landscape; lunar surface;
   moon landing; moon walk;
   rocks, 10, 20
   soil, 10, 20
   manned flights to, 8—10, 12, 20, 25, 26—28,
     60, 64
   unmanned flights to, 8—10, 12, 20
moon landing, 8—10, 25, 26—28, 60, 64
moon orbit, 25, 64
moon walk, 8—10, 26—28, 64

NASA, 57, 58, 60, 63, 64
Nixon, Richard M., 28

orbiters, 54, 57
O-rings, 54, 57

payload, 37, 48, 58

payload specialists, 38
physical fitness in space, 31
pilots, 38
Powers, John ("Shorty"), 18
Project Mercury: *see* Mercury Space Project

Reagan, Ronald, 47, 50, 54
remote manipulator systems (RMS), 38, 44
Ride, Sally, 42–45, 65
rocket boosters, 35, 54, 57
rockets: *see also* rocket boosters; space shuttle
    launching and, 14, 18, 25, 26
    moon orbit and, 26, 27
    lunar module and, 27, 28
    retro, 20

salvage missions, 46–48
*Salyut*, 65
satellites, communication: *see also* space stations,
    12, 13, 16, 23, 37, 38, 44, 64
    launchings, 44, 46, 48
    rescue of, 46–48
Saturn 5 rockets, 25, 26
Schmitt, Harrison, 28, 65
scientific experiments in space, 33, 38, 40, 60,
    63
scientists on space flights, 28
Shepard, Alan B., 17–18, 64
simulators, 41
*Skylab*, 30–33, 34, 58, 65
Soviet Union
    in space, 12, 14, 16–17, 22, 64, 65
    and space stations, 30
    and U.S. joint missions, 65
space, 10, 14
space colonies, 60
space flight: *see also* Mercury Space Project;
    Gemini Space Project; Apollo Space Project;
    space shuttle;
    manned, 8–10, 12, 20, 23–24, 25, 26–28,
    60, 64, 65
    unmanned, 8, 10, 12, 13, 14, 20, 64
space laboratories: *see* space stations, 48
space probes, 60, 63, 65
space shuttle, 13, 34–37, 38, 41, 42, 44–48,
    50, 54–57, 58–60, 65
    landings, 35, 37, 44
space stations, 13, 30–33, 34, 48, 65
    launching, 65
    manned, 58–60
space suits, 14, 18, 46, 57
space telescopes, 58

space walks, 24, 64
spacecraft, 14
*Spacelab*, 48, 58
*Sputnik*, 12, 13, 64

telephone technology, 13, 23
telescopes: *see* space telescopes
television transmission, 13, 23
Tereshkova, Valentina, 22, 64

United States in space: *see also* Gemini Space
    Project; Mercury Space Project; Apollo Space
    Project; space shuttle, 8–10, 12, 17–18,
    18–20
    and Soviet Union joint missions, 65
    and space stations, 30–33, 48

Venus, 65
*Viking* Space Probe, 60, 63, 65
*Vostok 1*, 16–17, 64
*Vostok 6*, 22, 64

Walker, David, 46
water immersion facility, 38–41
weightlessness, 14, 20, 31, 33, 38
White, Edward, 23–24, 64
women in space, 22–23, 42–45, 64, 65

Young, John, 37

zero gravity: *see* weightlessness

## PHOTO CREDITS

Front cover: NASA
All interior photographs courtesy of NASA except:
Pages 12, 13, 14 (top), 16, 17, 22 and 23 (top) from *Wide World.*
Pages 9, 15, 19, 36, 51, and 55 from *The New York Times.*
Page 23 (bottom) from Hughes Aircraft Company.